Deep-sea benthic foraminifera from Cretaceous–P boundary strata in the South Atlantic – taxonomy a

JOEN G.V. WIDMARK

Widmark, J.G.V. 1997 05 12: Deep-sea benthic foraminifera from Cretaceous–Paleogene boundary strata in the South Atlantic – taxonomy and paleoecology. *Fossils and Strata*, No. 43, pp. 1–94. Oslo. ISSN 0300-9491. ISBN 82-00-37667-2.

This study presents the results of a taxonomical investigation of deep-sea benthic foraminifera from Cretaceous–Paleogene boundary (K–PgB) strata at Deep Sea Drilling Project (DSDP) Sites 525 and 527 from the Walvis Ridge area, South Atlantic Ocean. Sites 525 (Walvis Ridge) and 527 (Angola Basin) represent different paleodepths, which were estimated to be about 1,100 and 2,700, respectively, at the time of the K–PgB event. A total of 36 samples were analyzed across the K–PgB; 12 samples are from the lowermost Danian and 24 from the uppermost Maastrichtian. The sections from the sites studied represent about 400 ka (about 200 ka before and after the K–Pg transition, respectively). The faunas at Sites 525 and 527 contain both calcareous and agglutinated benthic foraminifera, and they are dominated by the suborder Rotaliina. A total of 132 taxa were identified at the generic or specific level, and about 8% of the specimens encountered were not possible to identify, mostly because to their poor state of preservation. The benthic foraminiferal faunas at Sites 525 and 527 are very similar to the Paleocene 'Velasco-type' fauna (Berggren & Aubert 1975), which is characterized by, among others, *Gavelinella beccariiformis* (White), *Cibicidoides rubiginosus* (Cushman), *C. velascoensis* (Cushman), *Nuttallides truempyi* (Nuttall), *Nuttallinella florealis* (White), *Osangularia velascoensis* (Cushman), *Aragonia velascoensis* (Cushman), nodosariids (*Nodosaria velascoensis* Cushman, *Dentalina limbata* d'Orbigny), various agglutinated forms (*Gaudryina pyramidata* Cushman, *Tritaxia aspera* (Cushman), *Marssonella trinitatensis* Cushman & Renz), and various gyroidinoids and buliminids, of which most are 'relict' upper Maastrichtian species that survived the K–PgB. The Maastrichtian faunas at Sites 525 and 527 are dominated by *Gavelinella beccariiformis* (White) and *Nuttallides truempyi* (Nuttall), which maintained their dominance also in the lowermost Danian. The total faunal diversity (number of taxa) at the shallower Site 525 on the Walvis Ridge is somewhat higher (116 taxa) than the fauna (109 taxa) at the deeper Site 527 in the Angola Basin. The faunal difference between the two sites is probably paleobathymetrically controlled, and it is more pronounced in the Maastrichtian than in the Danian. Some Maastrichtian species that disappeared at the K–PgB [*Eouvigerina subsculptura* (McNeil & Caldwell), *Tritaxia aspera* (Cushman), *Loxostomum* sp., *Bolivinoides draco* (Marsson), *B. decoratus* (Jones), *Pseudouvigerina plummerae* Cushman, *Stensioeina pommerana* Brotzen, and *Sliteria varsoviensis* Gawor-Biedowa] occur at the shallower Site 525 only. On the other hand, four species that survived the K–PgB (*Globorotalites* sp. B, *Nuttallides* sp. A, *Nuttallides* sp. B, and *Pullenia* cf. *cretacea* Cushman) are common at the deeper Site 527 but absent at the shallower Site 525. With regard to the K–PgB event recently reported about in the literature, it is generally agreed that benthic foraminifera were not severely affected by this transition, compared to the mass mortality in other (planktic) marine organisms. Nevertheless, data in the literature suggest that benthic foraminiferal paleocommunities responded differently to this transition depending on paleobathymetric gradients, which influence interrelated ecological parameters such as the food flux to the sea floor and oxygen levels at the water–sediment interface and within the sediment itself. □*Benthic foraminifera, South Atlantic, Cretaceous, Tertiary, Cretaceous–Tertiary boundary, taxonomy, paleoecology, paleoenvironment, diversity, extinction.*

Joen G.V. Widmark [joen@gvc.gu.se], Department of Marine Geology, Earth Sciences Centre, Göteborg University, S-413 81 Göteborg, Sweden; 26 May, 1995; revised 30 March, 1996.

Contents

Introduction

The Walvis Ridge, situated in the eastern South Atlantic Ocean, forms a 2,000 km long and 200 km wide, roughly linear ridge extending in a northeasterly direction from the Mid-Atlantic Ridge to the African margin at about a latitude of 20°S (Fig. 1). Within the area of study the structural blocks slope steeply toward the Cape Basin, whereas in the northeast they slope more gradually toward the Angola Basin. The basement rocks range in age from 69 to 70 Ma, and the age of the basement decreases away from the ridge (Moore et al. 1983). The origin of the Walvis Ridge has been discussed for a long time, and a review of the geological and geophysical data published on the Walvis Ridge was provided by Dingle & Simpson (1976). Several hypotheses have been put forward in addressing the origin of the Walvis Ridge, including the 'hot-spot' hypothesis (Wilson 1965; Dietz & Holden 1970; Morgan 1972), the Heezen & Tharp (1965) 'micro-continent' hypothesis, the Ewing–Le Pichon uplift hypothesis (Ewing 1966; Zakharow 1981), the 'stress-release' hypothesis of Le Pichon & Hayes (1971), and finally, the 'sediment–basalt compensation during crustal separation' hypothesis (or the 'sea-floor-spreading hypothesis') by Shaffer (1984), which also is the one favored by Moore et al. (1983).

During Leg 74 of the Deep Sea Drilling Project (DSDP) in June–July, 1980, five sites (525–529) were drilled on the Walvis Ridge, southeastern Atlantic Ocean, along a NW–SE transect from the crest (at a depth of about 1,000 m) westward into the Angola Basin (4,400 m). The scientific purpose of this transect (about 230 km long) was to investigate the deep-water circulation in the southeastern

Atlantic, the nature and geologic evolution of the Walvis Ridge, and the biostratigraphy and magnetic stratigraphy of this region (Moore et al. 1983).

This study presents the results of a taxonomical investigation of the Upper Cretaceous – Lower Tertiary deep-sea benthic foraminifera from Sites 525 and 527 (Fig. 1). Previous studies of Cretaceous benthic foraminifera from the Walvis Ridge area have been made by Todd (1970), who analyzed Maastrichtian foraminifera from a deep-sea core drilled on the eastern flank of the Walvis Ridge, Beckmann (1978), who investigated Upper Cretaceous foraminifera from DSDP sites 363 and 364, and Scheibnerová (1978), who studied the same sites with respect to Aptian–Albian benthic foraminifera. The Paleogene benthic foraminifera from Walvis Ridge have been studied by Proto Decima & Bolli (1978) and Clark & Wright (1984). More recently, Widmark & Malmgren (1992a) studied benthic foraminiferal changes among the most common taxa (generally at the species level) across the Cretaceous–Paleogene (K–Pg) B at DSDP Sites 525 and 527. Several investigations on the relationships between benthic foraminiferal distribution and the paleoenvironment have been carried out on Late Cretaceous – Paleocene deep-sea material from the South Atlantic, e.g., by Ciesielski et al. (1977), Dailey (1983), Koutsoukos et al. (1990), Van Morkhoven et al. (1986), Parker et al. (1984), Scheibnerová (1971; 1972; 1973), Sliter (1976; 1977), Tjalsma & Lohmann (1983), Todd (1970), Widmark & Malmgren (1988, 1992b), and Widmark (1995). However, the Maastrichtian–Danian benthic foraminifera from Leg 74 have not been described in detail. The article on Cretaceous

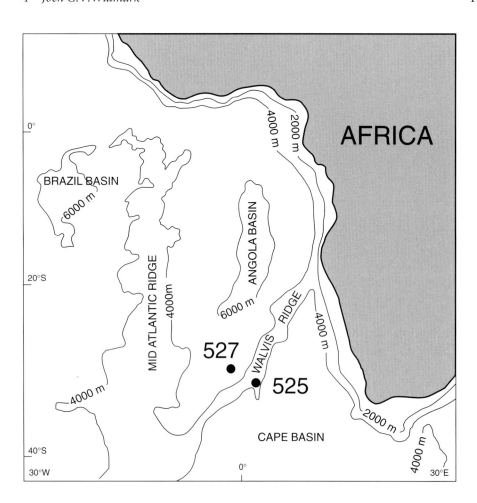

Fig. 1. Present-day location of DSDP Sites 525 and 527.

foraminifera in the Initial Report of the DSDP Leg 74 (Boersma 1984) concentrated on planktic foraminiferal biostratigraphy, and only the names of the most common benthic foraminiferal species in various intervals were given in the description of Site 525 in the Initial Reports of the DSDP Leg 74 (Moore *et al.* 1984a).

Acknowledgements. – I am very thankful to Brian Huber, Curator at the Smithsonian Institution (Washington, D.C.) and his wife Kathleen for their kindness and hospitality towards me and my wife and for providing me the opportunity to stay at the Smithsonian during a one-year post-doctoral scholarship supported by the Swedish Natural Science Research Council (NFR). Björn Malmgren (University of Göteborg) and Malcolm Hart (Polytechnic South West, Plymouth) are acknowledged for valuable comments on an early version of the manusscript, and so is Michael Kaminski (UCL, London) for kindly reviewing a later version. I am very grateful to Richard K. Olsson (Rutgers University, New Jersey) for his valuable comments on the final version of the manuscript and to Stefan Bengtson (Naturhistoriska Riksmuseét, Stockholm) for all his editorial input. Additional thanks go to Birgit Jansson, University of Uppsala, for preparing the samples and Christer Bäck, University of Uppsala, for his endurance in the dark room and for his photographic assistance in various ways. Deep-sea samples used in this study were generously supplied through the Ocean Drilling Program (ODP).

This study was supported by grants from the Th. Nordström and Hierta-Retzius Foundations, which were supplied through the Swedish Royal Academy of Sciences, and from the Swedish Natural Science Reserach Council (NFR). Publication costs were defrayed through NFR grant No. G-AA/GU 05239-310.

Material and methods

Benthic foraminifera across the K–PgB were analyzed from two DSDP Sites, 525 (Hole 525A) and 527 (Hole 527) in the Walvis Ridge area, South Atlantic. The 3.9 m section analyzed at Site 525 consists of a nannofossil chalk within Core 40 (Fig. 2); the K–Pg transition was encountered at the top of that core and is marked by a color change over an interval of about 15 cm (Sample 525A-40-2, 5–20 cm), from light brown to pale green (Moore *et al.*

Fig. 2. Lithology, magnetic polarity, and calcium carbonate at Hole 525A, Cores 30–63 (DSDP Site 525) (Modified after Moore *et al.* 1984a). The K–Pg boundary is situated just below the top of the reversed interval between Anomaly 29 and 30 in Core 40; all samples analyzed were obtained from that core (Table 2).

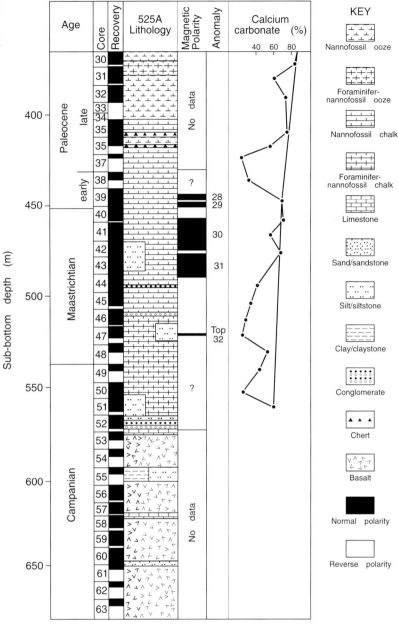

1984a). At Site 527 the studied interval is represented by a 5.71 m long section of a marly nannofossil chalk taken from Cores 32 and 33 (Fig. 3). The K–PgB itself is characterized by a sharp change from light to dark sediments in Sample 527-32-4, 50–52 cm (Moore *et al.* 1984b). At the time of deposition, the section in Site 525 was deposited at about 1,100 m (middle bathyal depths), whereas the one in Site 527 was considerably deeper (2,700 m), representing upper abyssal depths (Moore *et al.* 1984c).

A summary of the lithostratigraphy as well as the planktic foraminiferal and nannofossil biostratigraphies of Sites

525 and 527 were given by Moore *et al.* (1984a and b, respectively).

The magnetostratigraphy obtained at both sites facilitated estimates of the sedimentation rates for the intervals studied, which are situated within Subchron 29R (the reversed anomaly between Anomaly 29 and 30; Figs. 2 and 3). Sedimentation rates are here based on the assumption that the top of Anomaly 30 occurred about 0.3 Ma below the K–PgB and that the base of Anomaly 29 occurred about 0.3 Ma above the K–PgB (Berggren *et al.* 1985; Shackleton *et al.* 1984; Zachos & Arthur 1986)

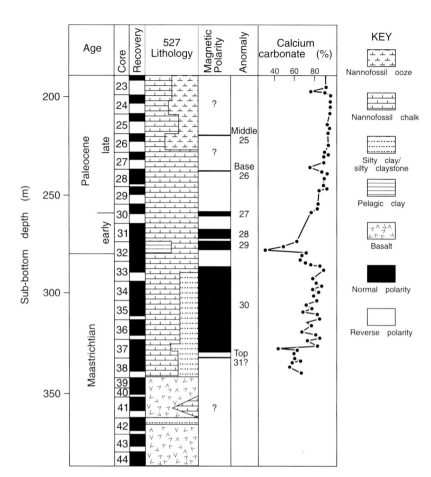

Fig. 3. Lithology, magnetic polarity, and calcium carbonate at Hole 527, Cores 23–44 (DSDP Site 527) (Modified after Moore *et al.* 1984b). The K–Pg boundary is situated just below the top of the reversed interval between Anomaly 29 and 30 in Core 32; samples analyzed were obtained from Cores 32 and 33 (Table 2).

(Table 1). The sections from the sites studied represent about 400 ka (about 200 ka before and after the K–Pg transition, respectively), within which a total of 36 samples from the two Sites 525 and 527 was analyzed (Table 2). Twelve samples are from the lowermost Danian (four and eight from Sites 525 and 527, respectively) and 24 (twelve from each site) are from the uppermost Maastrichtian. Data on depths of samples in site, estimated ages, sample weights, and total number of specimens of benthic foraminifera encountered in each sample are presented in Table 2. The Danian samples may be considerably younger, since 50–200 kyr of the basal Tertiary might be missing at these as well as at many other deep-sea cores, according to MacLeod & Keller (1991a,b), who based this conclusion on graphic correlation (MacLeod & Keller 1991b). However, Olsson & Liu (1993) were very skeptic to this conclusion and raised serious doubts against the reliability of the data of MacLeod & Keller (1991b), which implies that the question of 'incomplete deep-sea K–Pg-sections' is still open.

The samples were immersed in de-ionized water and placed on a rotary table for about 24 hrs. They were

Table 1. Data on present-day locations, water depths, paleolatitudes at about 60 Ma, paleodepths at the time of the K–PgB event, K–PgB depth, and estimated sedimentation rates for the DSDP holes used in the present study (see text for explanation). Figures within parentheses refer to sources of information.

Location	Hole 525A Walvis Ridge	Hole 527 Angola Basin
Latitude	29°S (2)	28°S (3)
Longitude	3°E (2)	2°E (3)
Water depth (m)	2,467 (2)	4,428 (3)
Paleolatitude	36°S (1)	36°S (1)
Paleodepth (m)	1,100 (4)	2,700 (4)
K–T boundary (Core-section, cm)	42-2, 11 (2)	32-4, 50 (3)
Depth below sea floor (m)	451.71	280.00
Base of Anomaly 29 (m above K–PgB)	0.7 (5)	2.0 (5)
Top of Anomaly 30 (m below K–PgB)	5.0 (5)	6.6 (5)
Estimated sedimentation rate above K–PgB (cm/ka)	0.2	0.7
Estimated sedimentation rate below K–PgB (cm/ka)	1.7	2.2

References: (1) Widmark & Malmgren (1992a); (2) Moore *et al.* (1984b); (3) Moore *et al.* (1984c); (4) Moore *et al.* (1984a); (5) Shackleton *et al.* (1984).

washed over a 63 μm sieve, and both fractions were dried and weighed. The sand fraction was then sieved over a 125 μm screen, and the larger fraction was used for collection of the benthic foraminiferal fauna and kept in slides for permanent reference. In most cases, the entire sample available was used. The number of specimens per sample (i.e. sample size) ranged between 108 and 657 (Table 2).

The taxonomy used here follows, with little exception, the one established by Loeblich & Tappan (1988). *Stensioeina* has coarsely perforated ventral sides and umbilical flaps, features that support its affinity to the genus *Gavelinella* and a placement within the subfamily Gavelinellinae; Loeblich & Tappan (1988) placed this genus within the subfamily Gyroidininae, of which most species of its type genus (*Gyroidinoides*) are characterized by smooth and finely-perforated umbilical (ventral) sides. The genus *Sliteria*, which is ventrally similar to *Stensioeina,* is here also placed in the subfamily Gavelinellinae on the basis of its ventral characteristics; it was originally described and placed together with *Stensioeina* within the subfamily Gyroidininae by Gawor-Biedowa (1992).

Examination of more than a hundred primary types (i.e. holotypes and paratypes) and secondary types (i.e. hypotypes, plesiotypes, and topotypes) from Upper Cretaceous – Lower Tertiary strata was carried out in the Cushman Collection, Smithsonian Institution, Washington, D.C. An account of the types examined is given in Appendix 1. In addition to the types examined, other collections, such as those of Todd (1970) and Todd & Low (1964) at the Smithsonian Institution and the Brotzen collection at the Naturhistoriska Riksmuséet in Stockholm was also consulted.

The taxa were imaged with a JEOL JSM 330 scanning microscope, using Ilford XP1-400 film; all illustrated specimens are deposited at the Department of Palaeozoology, Swedish Museum of Natural History, Stockholm, Sweden, under the numbers Pr2632–Pr2892.

Table 2. Data for DSDP samples used in the present study. Ages of samples below (-) and above (+) the K–Pg boundary were estimated using the chronology shown in Table 1. 'Weight' is total dry sample weight and N is the number of benthic foraminiferal specimens counted.

Sample No.	Site-Core-Sec.	Interval (cm)	Depth below seafloor (m)	Age (ka)	Weight (g)	N
1	525A-40-1	120–121	451.31	+200	3.77	413
2	525A-40-1	130–131	451.41	+150	3.34	436
3	525A-40-1	140–141	451.51	+100	4.09	330
4	525A-40-1	149–150	451.60	+60	2.40	560
	K–Pg boundary		451.71			
5	525A-40-2	47–48	452.08	20	6.08	350
6	525A-40-2	59–60	452.20	30	10.54	414
7	525A-40-2	69–70	452.30	40	7.63	497
8	525A-40-2	78–79	452.39	40	7.80	404
9	525A-40-2	92–93	452.53	50	4.41	560
10	525A-40-2	100–102	452.61	50	8.91	368
11	525A-40-2	109–110	452.70	60	5.57	499
12	525A-40-2	119–120	452.80	60	5.89	479
13	525A-40-2	128–129	452.89	70	6.22	295
14	525A-40-3	70–71	453.81	120	4.41	522
15	525A-40-3	130–131	454.41	160	6.67	657
16	525A-40-4	60–61	455.21	200	8.68	471
17	527-32-3	97–98	278.98	+150	2.07	165
18	527-32-3	117–118	279.18	+120	2.25	210
19	527-32-3	129–130	279.30	+100	3.61	331
20	527-32-3	139–140	279.40	+90	2.51	266
21	527-32-4	0–1	279.51	+70	4.17	489
22	527-32-4	19–20	279.70	+40	4.06	458
23	527-32-4	29–30	279.80	+30	3.67	239
24	527-32-4	39–40	279.90	+10	5.38	328
	K–Pg boundary		280.00			
25	527-32-4	57–59	280.08	<-10	12.43	509
26	527-324	68–69	280.19	10	6.50	240
27	527-32-4	87–88	280.38	20	6.41	178
28	527-32-4	107–108	280.58	30	6.16	170
29	527-32-4	133–134	280.83	40	5.60	167
30	527-32-5	7–8	281.08	50	6.25	156
31	527-32-5	33–34	281.33	60	5.61	208
32	527-32-5	57–58	281.58	70	5.50	163
33	527-32-5	92–94	281.93	90	17.69	298
34	527-32-6	30–31	282.81	130	5.40	108
35	527-32-6	91–92	283.42	160	6.08	132
36	527-33-1	18–20	284.69	210	8.22	234

Patterns of occurrence

The faunas at Sites 525 and 527 contain both calcareous and agglutinated benthic foraminifera, and they are dominated by the suborder Rotaliina. A total of 12,411 specimens were encountered; 92% of these are identified and referred to 132 taxa at the generic or specific level. The rest, about 8% of the specimens, were not possible to identify, mostly because of the poor state of preservation. Of the 132 taxa identified, 55 taxa (56% of the specimens) are described using 'closed' nomenclature, 12 taxa (5% of the specimens) are compared (using 'cf.') with species

documented by other workers, 29 taxa (18% of the specimens) are left in open nomenclature, and, finally, 36 taxa (15% of the specimens) were lumped at the generic level. New species were deliberately not established in this communication, because I believe that there is already a redundancy of Upper Cretaceous benthic foraminiferal species described in the literature, many of which should be regarded as synonyms. The establishment of additional species that might already have been described would undoubtedly increase the confusion in Upper Cretaceous benthic foraminiferal taxonomy. Range charts showing the absolute abundances of the taxa and arranged according to their last appearances are given in Appendix 2.

The Maastrichtian faunas at Sites 525 and 527 are dominated by *Gavelinella beccariiformis* (White) and *Nuttallides truempyi* (Nuttall). They maintained dominance also in the lowermost Danian, indicating quite well-oxygenated conditions throughout the intervals studied. The total faunal diversity (number of taxa) at the shallower Site 525 on the Walvis Ridge is somewhat higher (116 taxa) than for the fauna (109 taxa) at the deeper Site 527 in the Angola Basin. The faunal difference is probably paleobathymetrically dependent and controlled by ecological gradients, such as decreasing food availability and increasing oxygen levels, from shallower to deeper environments. The difference in faunal composition and fanal diversity between the two sites is more pronounced in the Maastrichtian than in the Danian. Some Maastrichtian species that disappeared at the K–PgB [*Eouvigerina subsculptura* (McNeil & Caldwell), *Tritaxia aspera* (Cushman), *Loxostomum* sp., *Bolivinoides draco* (Marsson), *B. decoratus* (Jones), *Pseudouvigerina plummerae* Cushman, *Stensioeina pommerana* Brotzen, and *Sliteria varsoviensis* Gawor-Biedowa] occur at the shallower Site 525 only. On the other hand, four species that survived the K–PgB (*Globorotalites* sp. B, *Nuttallides* sp. A, *Nuttallides* sp. B, and *Pullenia* cf. *cretacea* Cushman) are common at the deeper Site 527 but absent at the shallower Site 525.

The benthic foraminiferal faunas at Sites 525 and 527 are very similar to the Paleocene 'Velasco-type' fauna, which was defined by Berggren & Aubert (1975). The 'Velasco-type' fauna is characterized by, among others, *Gavelinella beccariiformis* (White), *G. rubiginosa* (Cushman), *G. velascoensis* (Cushman) [= *Cibicidoides velascoensis* (Cushman)], *Nuttallides truempyi* (Nuttall), *Nuttallinella florealis* (White), *Osangularia velascoensis* (Cushman), *Aragonia velascoensis* (Cushman), nodosariids (*Nodosaria velascoensis* Cushman, *Dentalina limbata* d'Orbigny), various agglutinated forms [*Gaudryina pyramidata* Cushman, *Tritaxia aspera* (Cushman), *Marssonella trinitatensis* Cushman & Renz], and various gyroidinoids and buliminids (Berggren & Aubert 1975), of which most are 'relict' upper Maastrichtian species that survived the K–PgB (Tjalsma & Lohmann 1983). The 'Velasco-type' fauna has previously been described and illustrated in several publications, e.g., on the Velasco Formation of Mexico by Cushman (1926b), White (1928a, b, 1929), and Cushman & Renz (1946), and on Maastrichtian and Paleocene DSDP/ODP material by Webb (1973), Sliter (1976, 1977), Beckmann (1978), Dailey, (1983), Tjalsma & Lohmann (1983), Widmark & Malmgren (1988, 1992a, b), Thomas (1990), and Nomura (1991). The 'Velasco-type' fauna has, from general geologic considerations, been regarded as representing deepwater environments (Saint-Marc 1987). *Nuttallides*

truempyi (Nuttall), one of the dominating species in most of the samples analyzed here, was suggested by Berggren & Aubert (1983) to inhabit an upper depth limit of 500-600 m. This agrees with the estimated paleodepths given by Moore *et al.* (1983c) for Sites 525 and 527 (1,100 and 2,700 m, respectively).

Faunal changes across the K–PgB

Several quantitative investigations on benthic foraminiferal changes across the K–PgB have emerged during the last years based on material representing a wide paleolatitudinal as well as paleobathymetrical range. The benthic foraminiferal response to the K–PgB event in neritic to middle-bathyal environments have been investigated from low-latitude landbased sections from Texas (Brazos River; Keller 1989), Tunisia (El Kef; Keller 1988, 1992; Speijer & Van der Zwaan 1996), Middle East (Negev–Sinai Border; Keller 1992), Southern Spain (Caravaca; Keller 1992; Coccioni & Galeotti 1994), and Northern Spain (Basque Basin; Kuhnt & Kaminski 1993); a mid-latitude, upper bathyal land based section was studied by Kaiho (1992) from northern Japan (Kawaruppu section, Hokkaido Island). Several K–PgB sections from DSDP/ODP cores have also been quantified with respect to benthic foraminiferal changes across this transition. Studies concerning the middle-bathyal–upper-abyssal environment include investigations on the low-latitude Central Pacific Site 465 (Hess Rise; Widmark & Malmgren 1992a), the mid-latitude South Atlantic Sites 516, 525, and 527 (Rio Grande Rise – Walvis Ridge; Dailey 1983; Widmark & Malmgren 1992a), and high-latitude sites from the Indian (Site 752, Broken Ridge; Nomura 1991) and Southern Ocean (Sites 689 and 690, Maud Rise; Thomas 1990). A listing of these previous quantified K–Pg sections is given in Table 3 including the size fraction analyzed and their paleolatitude and depth of deposition at the time of the K–PgB event; their paleolocations are given on paleogeographic reconstruction in Fig. 4.

Faunal turnover (diversity changes and 'extinctions' across the Cretaceous–Paleogene boundary)

The most usual way of presenting faunal changes is in terms of 'faunal turnover', which is often referred to as species (taxon) 'extinction' or disappearance (appearance) during a certain interval of (geological) time. Faunal turnover patterns obtained from the studies mentioned above will be reviewed and discussed in the following.

Keller (1992) compared landbased K–PgB sections from the Tethyan margin including the El Kef, Brazos

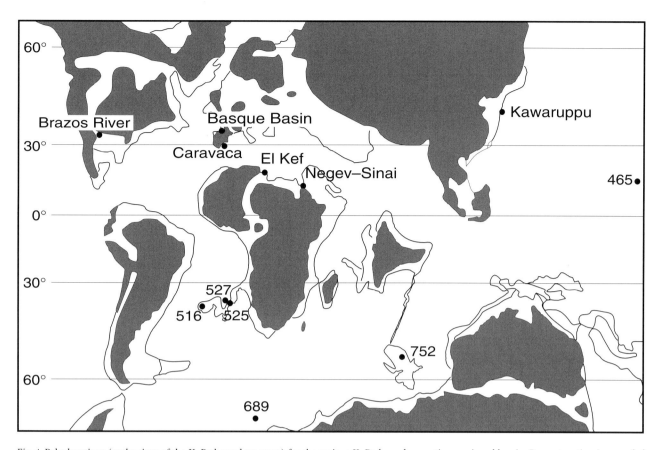

Fig. 4. Paleolocations (at the time of the K–Pg boundary event) for the various K–Pg boundary sections reviewed herein. Reconstruction is compiled from various sources: North America, Europe, North Africa, northern South America (Keller 1992); South Atlantic (Shaffer 1984); Indian Ocean (Klootwijk *et al.* 1992); Japan (Kaiho 1992).

Table 3. Sections previously studied with respect to benthic foraminiferal changes across the K–Pg boundary (see also Fig. 4).

Location of section analyzed	Size fraction (μm)	Paleolatitude	Paleobathymetry	Author
Brazos River, Texas	106	30°N	o. neritic – m. neritic	Keller (1989a, 1992)
El Kef, Tunisia	150	20°N	u. bathyal – o. neritic	Keller (1988, 1992)
El Kef, Tunisia	125	20°N	u. bathyal – o. neritic	Speijer & Van der Zwaan (1996)
Negev–Sinai Border, Middle East	106	15°N	u. bathyal – o. neritic	Keller (1992)
Caravaca, southern Spain	106	30°N	u. bathyal – o. neritic	Keller (1992)
Caravaca, southern Spain	125	30°N	u. bathyal – o. neritic	Coccioni & Galeotti (1994)
Basque Basin, northern Spain	not given	35°N	middle bathyal	Kuhnt & Kaminski (1993)
Hokkaido Island, Japan	63	50°N	upper bathyal	Kaiho (1992)
Hess Rise, Pacific Ocean (DSDP Site 465)	125	16°N	middle bathyal	Widmark & Malmgren (1992a)
Rio Grande Rise, South Atlantic (DSDP Site 516)	150	36°S	middle bathyal	Dailey (1983)
Walvis Ridge, South Atlantic (DSDP Site 525)	125	36°S	middle bathyal	Widmark & Malmgren (1992a)
Walvis Ridge, South Atlantic (DSDP Site 527)	125	36°S	upper abyssal	Widmark & Malmgren (1992a)
Broken Ridge, Indian Ocean (ODP Site 752)	149	50°S	middle bathyal	Nomura (1991)
Maud Rise, Southern Ocean (ODP Site 689)	63	70°S	middle bathyal	Thomas (1990)
Maud Rise, Southern Ocean (ODP Site 690)	63	70°S	lower bathyal	Thomas (1990)

River, Negev–Sinai Border, and Caravaca sections, of which the first two initially were studied by Keller (1988 and 1989a, b, respectively). At the deeper, upper-bathyal–outer-neritic sections (i.e. the El Kef, Negev–Sinai Border, and Caravaca), 39–44% of the Cretaceous species were restricted to or dominant (rare in the Tertiary) only in the Maastrichtian, 20–32% of the Tertiary species were restricted to or dominant (rare in the Cretaceous) only in the early Paleocene, and, finally, the species ranging across the K–PgB were found to constitute 29–36% of the Cretaceous fauna (Keller 1992). The shallower, middle- to outer-neritic Brazos sections were quite different from the other sections mentioned above in showing only about half the species diversity but, in spite of that, a stronger faunal turnover across the K–PgB: as much as 50% of the Cretaceous species were restricted to or dominant in the Maastrichtian (rare in the Tertiary), comparably more of the Tertiary species (33%) were restricted to or dominant (rare in the Cretaceous) in the basal Paleocene, leading to the fact that only two species (17%) of the Cretaceous species ranged across the K–PgB and survived at Brazos River (Keller 1992). However, these data are somewhat difficult to evaluate since Keller (1989b, 1992) placed the K–PgB considerably higher than in other studies on the general stratigraphy in the Gulf Coastal Plain, to which the Brazos River succession belongs; the faunal turnover will thus be different depending on where the boundary is placed. The argument concerns the age of the 'tsunami beds', which are placed just *above* the K–PgB by a number of workers (eg., Montgomery *et al.* 1992; Smit *et al.* 1994; Smit 1995; Pospichal 1995; Rudolph *et al.* 1995), in various sections in the Gulf Coastal Plain. Keller (1989b, 1992), on the other hand, placed the tsunami layer several beds *below* the boundary at Brazos River and it might be assumed that the high turnover rate of Keller (1992) in fact is an artifact, especially since Montgomery *et al.* (1992) had strong arguments to consider the Brazos River succession incomplete owing to the absence of the *Abathomphalus mayaroensis* Zone.

Speijer & Van der Zwaan (1996) argued for a quite abrupt faunal turnover at the K–PgB at El Kef in recording an (at least local) extinction of 39% of the latest Maastrichtian taxa within the uppermost meter of the Cretaceous (excluding the lumped taxa on the suprageneric level). Based on the 60 most common taxa in the sequence, Speijer & Van der Zwaan (1996) identified six stratigraphic assemblages, each with a specific range, which show three steps in benthic foraminiferal development. These steps coincided with the Maastrichtian (uppermost *Abathomphalus mayaroesis* Zone), the earliest Paleocene (the three biozones of *Guembelitria cretacea, Parasubbotina fringa,* and *Parvularugoglobigerina eugubina*), and the early Paleocene (*Parasubbotina pseudobulloides* Zone) (Speijer & Van der Zwaan 1996).

Coccioni & Galeotti (1994) made a detailed study (on the millimeter/millennial scale) across the K–PgB at Caravaca, southeastern Spain. Since they did not identify the taxa on species but genus level, no information on species diversity or disappearances was presented. Nevertheless, a drastic drop in genus richness, from about 25 genera in the upper Maastrichtian down to only two surviving genera observed in the second boundary-clay sample (the first sample above the boundary was barren of foraminifera) was recorded (Coccioni & Galeotti 1994); in the following boundary-clay samples there was a steady increase in genus richness, which was back to Maastrichtian values just above the boundary clay.

Kuhnt & Kaminski (1993) quantified faunal changes across the K–PgB in deep-water agglutinated foraminifera (DWAF) from the uppermost Maastrichtian through the boundary clay in two mid-latitude landbased sections (Zumaya and Sopelana) in the Basque Basin, northern Spain. They concluded that no major extinction among DWAF took place across the K–PgB. Out of 45 DWAF species observed by Kuhnt & Kaminski (1993) in the uppermost Maastrichtian in both sections, nine (20%) and ten (22%) were not observed in the boundary clay above the K–PgB in the Zumaya and Sopelana sections, respectively; this minor loss in species richness was almost completely compensated for by 'new' species that emerged above the boundary.

Kaiho (1992) studied benthic foraminiferal changes across the mid-latitude K–PgB in a section deposited in an upper bathyal environment at Kawaruppu, Hokkaido Island, Japan. Out of the 40 Maastrichtian species, only ten (25%) were absent in the Danian, in which eight 'new' species appeared (Kaiho 1992), resulting in a nonsignificant diversity loss of two species across this K–PgB. Kaiho (1992), however, concluded that five of the species absent in the Danian section have actually been reported from Tertiary strata elsewhere and that they, therefore, could not be considered extinct, resulting in an 'extinction rate' of 13% only. Furthermore, if only the calcareous benthics were taken into account, the 'extinction rate' would be as low as 10% (Kaiho 1992).

Dailey (1983) quantified mid-latitudinal middle bathyal benthic foraminifera from the Santonian through late Paleocene at Site 516 (Hole 516F) from the Rio Grande Rise (South Atlantic) and laid a certain emphasis on their faunal response to the K–PgB event. He reported an accelerated rate of disappearances toward the boundary that culminated in the final 0.5 Ma of the Maastrichtian. During this interval, 28% of the Maastrichtian species appeared for the last time and did not survive the K–PgB (Dailey 1983). In spite of the appearance of some 'new' species in the first 0.5 Ma interval above the K–PgB, there was a net loss in species diversity from about 75 species in the Maastrichtian to about 60 species in the basal Danian (Dailey 1983).

Widmark & Malmgren (1992a) investigated benthic foraminiferal changes across the K–PgB in the low-latitudinal, middle-bathyal, Pacific (Hess Rise; DSDP Site 465) and South Atlantic (Walvis Ridge; DSDP Sites 525 and 527), of which the two latter (mid-latitudinal, middle-bathyal–upper-abyssal) sites are taxonomically accounted for in the present communication. In determinating species (taxon) disappearances and appearances in the Danian, Widmark & Malmgren (1992a) used both 'minimum' and 'maximum' disappearances–appearances, where the 'minimum' values were based upon species (taxa) that occurred in at least 50% of the Maastrichtian or Danian samples, and the 'maximum' values included also the rarer species (taxa). They found that between 5–13% and 23–40% of the taxa encountered in the Maastrichtian samples disappeared above the K–PgB and that the highest taxonomic disappearance (23–40%) was observed in the shallower South Atlantic mid-latitude fauna (Site 525), whereas the lowest (5–13%) was in the Pacific low-latitude fauna (Site 465). The deeper South Atlantic mid-latitude fauna (Site 527) showed an intermediate percentage of 9–18%, implying that the taxonomic disappearance is about two times higher at shallower (middle bathyal) depths than at deeper (lower-bathyal to abyssal) depths in the South Atlantic (Widmark & Malmgren 1992a). In addition, these authors found that only between 3–5% and 3–12% of the lowermost Danian benthic foraminiferal fauna originated in the lowermost Danian and that this did not compensate for the loss of Maastrichtian taxa across the K–PgB. Since taxonomic diversities are dependent upon sample size (larger samples generally contain more species than smaller samples; e.g., Kempton 1979; Parker *et al.* 1984; Malmgren & Sigaroodi 1985), Widmark & Malmgren (1992a) standardized the diversity to the number of taxa for a sample size of 100 specimens using Hurlbert's diversity index (Hurlbert 1971). They recorded a simultaneous diversity increase at both the South Atlantic sites within the interval 101–150 ka below the K–PgB, from 29 to 39 and from 26 to 40 taxa at Sites 525 and 527, respectively. At Site 465, diversity changed from 23 taxa in the 151–200 ka interval to a maximum of 27 taxa in the latest Maastrichtian (0–50 ka before the K–PgB). During the first 150 ka of the lowermost Danian, diversity decreased at all three sites to 28, 34 and 18 taxa at Sites 525, 527 and 465, respectively. However, the extremely low taxonomic disappearance (5–13%) documented by Widmark & Malmgren (1992a) from Site 465 is probably an artifact due to the 'mixed zone' of Danian and Maastrichtian sediments that is smeared throughout about 30 cm of the basal Danian in Hole 465A (Thiede *et al.* 1981). The three lowermost Danian samples in Widmark & Malmgren's (1992a) data set from Site 465 are within this 'mixed zone', which most probably lagged the faunal response to the boundary event and thus generate a low taxonomic

disappearance. In order to test this, the data from Site 465 of Widmark & Malmgren (1992a) was recalculated for more reliable minimum and maximum taxonomic disappearances by excluding the three 'mixed-zone' samples. The recalculation resulted indeed in much higher taxonomic disappearances for Site 465 in ranging between 22% and 44%, which is comparable with mid-latitude values from similar paleodepths (Site 525; 23–40%).

Nomura (1991) studied the paleoceanography of upper Maastrichtian – Eocene benthic foraminifera at three ODP sites (752, 753 and 754) from Broken Ridge in eastern Indian Ocean, and of which Site 752 (Hole 752B) yielded a continuous section across the K–PgB. For calculations of species disappearances across the K–PgB, Nomura (1991) considered all 187 upper Maastrichtian – Paleocene taxa (including rare species) and concluded that out of these 187 taxa, only 23 (12.3%) did not survive into the Danian.

Thomas (1990) analyzed Late Cretaceous through Neogene deep-sea benthic foraminifera from two high-latitude ODP sites (689 and 690) from Maud Rise in the Weddell Sea (Southern Ocean). She found no major extinctions among these organisms and reported that only 14% and 9% of the Maastrichtian species at Sites 689 and 690, respectively, disappeared (globally or locally) within an interval of 0.5 Ma above and below the K–PgB. This implies that Thomas (1990) found a higher disappearance in the faunas representing shallower, middle bathyal (Site 689) than deeper, lower bathyal (Site 690) environments.

On the basis of the results reviewed above, the general pattern emerges that benthic foraminiferal faunas were not severely affected by the K–PgB event compared to the mass mortality in, e.g., nannofossils and planktic foraminifera. However, Widmark & Malmgren (1992a) demonstrated a clear faunal response to the K–PgB event by using correspondence analysis, which clearly separated the Danian faunas (samples) from the ones of Maastrichtian age. They concluded that most of the species included in their correspondence analyses responded to the event in terms of increases or decreases in abundance, whereas only a few disappeared at the K–PgB itself.

Benthic foraminiferal ecology and morphotypic changes across the K–PgB

The biogeographic and ecologic (bathymetric) distribution of fossil as well as Recent deep-sea benthic foraminifera is controlled by the environment they live in, which in turn is determined by several physical, chemical, and biological parameters. Some parameters, such as salinity and temperature, are mainly dependent on the surrounding water mass, whereas others, such as food and oxygen levels, may vary considerably within a given water mass or oceanic basin because of local upwelling and/or the (seasonal) variation of primary production rates.

During the last decade, several papers have demonstrated that individual species of Recent deep-sea benthic foraminifera occupy different microhabitats, i.e. that they live at different depths in relation to the water–sediment interface (e.g., Corliss 1985, 1991; Gooday 1986; Mackensen & Douglas 1989). It has also been shown that there is a relationship between the microhabitat and the morphology of the foraminiferal test: species found mainly on top of the sediment or within the top 0.5 cm are widely referred to as epifaunal and dominated by trochospiral test morphologies, whereas other species, which are found deeper down in the sediment, are referred to as infaunal and exhibit a variety of bi-, tri-, and multiserial and planispiral test morphologies (Corliss 1985, 1991). It has been shown that generally benthic foraminiferal densities, as well as the densities and microhabitat depth of infaunal species, are linked to the trophic structure (i.e. food and/or oxygen levels) of the environment where they live. Which of the two (food or oxygen) is the limiting factor on benthic foraminiferal densities and microhabitat depths has been under dispute for quite some time: some workers (e.g., Mackensen & Douglas 1989; Barmawidjaja *et al.* 1992; Jorissen *et al.* 1993; Kaminski *et al.* 1995) found that oxygen was the limiting factor affecting the benthic foraminiferal densities, morphogroups, and microhabitat depths in their material, whereas others (e.g., Gooday 1986; Corliss & Emerson 1990) attributed both oxygen and food more or less equal importance as controlling factors on benthic foraminiferal communities. Finally, there is a third standpoint based on high frequencies of infaunal morphotypes found in connection with high organic-carbon fluxes, suggesting that food is the most important factor controlling the ecology of and morphotypic distribution of benthic foraminifera (Corliss & Chen 1988; Corliss 1991; Rosoff & Corliss 1992; Rathburn & Corliss 1994). However, in spite of the contradiction on this matter throughout the literature, it is generally accepted that epifaunal species are less tolerant to low-oxygen levels than infaunal species, which obviously flourish in environments where the food-supply is high and the oxygen level is low (because of the oxygenation of organic matter). Infaunal species are thus opportunistic (*r*-selected) and are able to reproduce at high turnover rates in order to profit from large food resources, whereas epifaunal species in general are equilibrium species (*K*-selected), which implies that they are able to sustain themselves under conditions with quite low food supply as long as the oxygen levels are kept at reasonably high levels. However, high frequencies of some Recent epifaunal species (i.e. *Epistominella exigua* and *Alabaminella weddellensis*) have also been linked to sudden (seasonal) high inputs of phytodetritus (Gooday 1988, 1994; Gooday & Lambshead 1989) and they should

thus be regarded as opportunistic (*r*-selected) species. This implies that their distribution actually is controlled by large organic-carbon inputs that are quite restricted temporally and spatially; this of course questions the general usage of epifaunal species as deep-water tracers (e.g., Corliss 1985; Thomas & Vincent 1987, 1988; Linke & Lutze 1993).

In summary, generally high frequencies of infaunal morphotypes indicate eutrophic conditions with high organic-carbon fluxes and low oxygen levels. The absence of infaunal morphotypes generally indicates opposite, oligotrophic (food-poor) and well-oxygenated conditions, under which the remaining fauna (i.e. largely epifaunal morphotypes) probably shows a relation with the surrounding water-mass characteristics. As soon as there is a significant change (or variation) in the trophic regime, then this relationship is overruled by the availability of food, which is a much stronger ecological factor than the physicochemical characteristics of the surrounding water mass.

There is always a problem with direct comparisons of benthic foraminiferal (and especially fossil) species distribution data of different origin because of incompatible taxonomic species concepts applied throughout the literature. The conceptual relationship between benthic foraminiferal morphotypes and environmental parameters may, however, provide a tool for interpretations of paleoenvironmental changes in the geological past if we assume that fossil morphotypes have the same, or at least similar, ecological requirements as their Recent morphological counterparts, of which the ecological requirements are (fairly) well known.

This approach has been used by several workers in applying the model established by Corliss & Chen (1988). Most of these workers (i.e. Thomas 1990; Keller 1992; Widmark & Malmgren 1992a) recorded a conspicuous decline in relative abundances by the infaunal morphotypes across the K–PgB, indicating that the opportunistic (*r*-selected) component, which is dependent upon high-food and/or low-oxygen levels, suffered much more from the K–PgB event than the epifaunal (*K*-selected) component of the benthic community. A depth-related trend was also found to affect this change in benthic community structure. Keller (1992) concluded that infaunal habitats in the upper bathyal were most severely affected but that the benthic community as a whole suffered more in shallower, mid-neritic environments. In the deep-sea environment, Widmark & Malmgren (1992a) observed an infaunal decline of 49 percentage units in the middle bathyal environment (Site 525), from 64% to 15% across the boundary, whereas in the deeper and more oligotrophic, upper abyssal environment at Site 527, the infaunal decline across the K–PgB was only about 10 percentage units, from 24% to 14%. However, data from the high resolution study by Coccioni & Galeotti (1994) show a contradictory pattern in recording a significant infaunal *increase* immediately above the K–PgB at Caravaca where a single infaunal morphogroup consisting of *Bolivina* and *Spiroplectammina* bloomed opportunistically.

The causes for these benthic community changes across the K–PgB have been interpreted differently in the various studies reviewed here. For instance, Keller (1992) linked the decline in infaunal abundances to low oxygen levels due to the early Danian sea-level transgression that should result in reduced upwelling, an enhanced oxygen-minimum zone, and a suppressed primary production in neritic–upper-bathyal environments. However, Keller's (1992) view is contradicted by the one of Kaiho (1992), who established an oxygen index, which indicated quite well-oxygenated conditions in mid-latitude upper-bathyal environments in the western Pacific at the time of the K–PgB event. Kaiho (1992) suggested that the break-down of the marine food-chain instead was induced by darkness and acid rains initially triggered by an asteroidal impact at the time for the K–PgB event; such a consequence of an asteroidal impact has earlier been proposed by, e.g., Lewis *et al.* (1982) or Prinn & Fegley (1987).

Coccioni & Galeotti (1994), as well as Speijer & Van der Zwaan (1996), offered alternative scenarios including both a collapse of the primary productivity at the surface *and* low oxygen levels at the K–PgB. The former two authors argued that the plankton mass-mortality at the surface caused an extensive input of organic matter (so-called 'nutrient soup') to the sea floor at Caravaca as a result of a catastrophic and (geologically) instantaneous event. Such an amount of a sudden influx would lower the oxygen levels through degradation and kill off most of the benthics but a few opportunistic species that are able to profit from large food resources and that are tolerant to anoxic environments (Coccioni & Galeotti 1994). Speijer & Van der Zwaan (1996) suggested that at El Kef the latest Maastrichtian was characterized by high fertility due to moderate upwelling, which was driven by the formation of warm, saline intermediate waters within the Tethyan Realm and by the strong trade winds, which were blowing along the shelves of the southern Tethyan margin. These conditions changed because of a cooling event at the K–PgB that resulted in a blocking of warm, saline deep water formation and weakening trade winds (because of weaker atmospheric contrast), which in turn led to decreasing upwelling and primary productivity (Speijer & Van der Zwaan 1996). Furthermore, Speijer & Van der Zwaan (1996) argued that the near-cessation of warm, saline intermediate water replenishment resulted in a strongly developed OMZ and low-oxygen conditions (despite decreased primary productivity) and that the whole scenario could have been trigged by an asteroidal impact at the time of the K–PgB. However, there is no evidence that low-oxygen conditions at the K–PgB should have been a global phenomenon, and indeed, benthic foraminiferal communities from the deep-sea instead indicate quite well-oxygenated conditions in the deep-sea environment (e.g., Dailey 1983; Thomas 1990; Nomura 1991; Widmark & Malmgren 1992a).

Thomas (1990) and Widmark & Malmgren (1992a) suggested that the decrease in the infaunal population was related to a drastic decline in the surface-water productivity, which resulted from the collapse of the food-web and in a lowering of the organic-carbon flux to the deep-sea environment; a similar scenario was proposed by Kuhnt & Kaminski (1993) based on changes in the community structure among middle-bathyal DWAF. Widmark & Malmgren (1992a) were of the opinion that both the asteroid impact hypothesis (Alvarez *et al.* 1980, 1984) and the one advocating intense volcanism (McLean 1981, 1985; Officer & Drake 1983, 1985; Officer *et al.* 1987) are compatible with a scenario involving the break-down of primary productivity in the photic zone that resulted in a decline of the organic-carbon flux to the deep-sea. Kuhnt & Kaminski (1993) did not exclude the possibility of a single impact event to trigger the biotic crisis at the K–PgB, but the comparable long-term recovery period (see below) found in their DWAF data could also, according to these authors, imply a more steady and persistent source of environmental deterioration such as unusually high CO_2 levels in the atmosphere.

Recovery of benthic communities in the post-K–PgB environment

The recovery time for the benthic foraminiferal community after the biotic crisis at the K–PgB has been shown to vary considerably. For instance, Keller (1992) suggested that environmental conditions at El Kef were back to normal about 250–300 ka after the boundary in the uppermost part of Zone P1a, when carbonate sedimentation increased and with the reappearance of Lazarus taxa and the expansion of the infaunal community. A similar duration was recorded by Speijer & Van der Zwaan (1996), who discriminated six assemblages at El Kef that responded differently to the K–PgB event, of the Lazarus taxa (assemblage SA2 that disappeared at the boundary), which also reappeared in the upper part of the *P. eugubina* Zone (= P1a of Keller 1992) indicating more stabilized environmental conditions at that time. Coccioni & Galeotti (1994) calculated with a much faster recovery of the benthic ecosystem in the similar environment at Caravaca. They recorded suppressed environmental conditions only through Zone P0 (the boundary clay) and normal carbonate accumulation rates already at the base of Zone P1a and concluded that aerobic, normal conditions were achieved about 7 ka after the boundary by the reappearance of polytaxic (*K*-selected) assemblages. Kuhnt & Kaminski (1993) found a slow recovery by middle bathyal

DWAF that took probably 50–100 ka; this was based on the observation of agglutinated assemblages using calcareous cement in the upper part of the boundary clay (Zone P0). Kuhnt & Kaminski (1993) and Coccioni & Galeotti (1994) interpreted the boundary clay (Zone P0) in the Basque Basin and Caravaca, respectively, to represent suppressed, low-oxygen conditions, after which the benthic community almost recovered. The time span during which the boundary clays at Caravaca and in the Basque Basin were deposited seems, however, to vary considerably, because the former study was based on the biochronostratigraphy established by Berggren & Miller (1988) and the latter on that of Smit (1990). At El Kef the stressed environment seems to have lasted much longer compared to the deeper environments of Caravaca and the Basque Basin and ended first at the top of Zone P1a by the recovery of the infaunal component of the foraminiferal benthic fauna (Keller 1992). However, comparisons between recovery times of the studies reviewed above become somewhat confused because of the lack of standardization concerning the duration of the biostratigraphic zones used for the Danian sections. Furthermore, Danian zone designations are used quite differently, in that Zone Pα of Berggren & Miller (1988) is the same as Zone P1a of Keller (1992) and Zone P1a of Berggren & Miller (1988) is equal to Zone P1b of Keller (1992).

Unfortunately, benthic foraminiferal recovery times in connection with the K–PgB event in the deep-sea environment from the DSDP/ODP cores reviewed earlier are hard to evaluate. This is due to the quite low-resolution sampling procedures used by Dailey (1984), Thomas (1990), and Nomura (1991) and because the Danian interval analyzed by Widmark & Malmgren (1992a) was too short for detecting the reappearance of Lazarus taxa and other faunal events further up in the Paleocene. Another problem with K–Pg-boundary studies on deep-sea material concerns the possibility that the lowermost Danian (Zones P0–P1a) is lacking in most deep-sea sections, which should imply a 50–200 ka long period of nondeposition in the earliest Paleocene (MacLeod & Keller 1991a,b); this has, however, been heavily disputed by Olsson & Liu (1993).

Finally, it must be kept in mind that the model of, e.g., Corliss & Chen (1988) is somewhat simplified and that an uncritical employment of such models may lead to erroneous interpretations of the environmental conditions especially when applied on fossil assemblages. One example of this concerns the general use of trochospiral, epifaunal taxa as indicators of more oligotrophic conditions (i.e. deep-water characteristics), but, as mentioned before, some Recent epifaunal species have indeed been shown to be able to profit from large inputs of phytodetritus and thus indicate more eutrophic conditions. Another example involves the decline and recovery by the infaunal component, in which some infaunal Cretaceous taxa responded quite differently to the K–PgB event. At

the middle bathyal South Atlantic Site 525 the Cretaceous fauna is marked by abundant infaunal *Praebulimina* spp. and *Eouvigerina subsculptura* along with dominating epifaunal *Gavelinella beccariiformis* and *Nuttallides truempyi* (Widmark & Malmgren 1992a). The former two (infaunal) species disappeared across the K–PgB, and this was used by Widmark & Malmgren (1992a) as evidence for a breakdown of the stable primary-productivity conditions at the end of the Cretaceous. Simultaneous with the disappearance of these taxa, there was a conspicuous increase of some other infaunal species represented by *Bulimina spinea* (usually referred to as *B. midwayensis*; see systematic taxonomy, this volume), *B. trinitatensis*, and *Aragonia* spp., which obviously did not respond to the suppressed primary productivity at the K–PgB as may be expected from 'eutrophic' (infaunal) species. The same pattern is even better expressed at the middle bathyal Pacific Site 465, where *Praebulimina* spp. disappear in the lowermost Danian, whereas, similar to Site 525, *B. spinea* and *B. trinitatensis* increase in relative abundances together with some other infaunal species such as *Pyramidina rudita* and *Quadratobuliminella* spp.; in addition, there are stable frequencies of *Aragonia* spp. and *B. velascoensis* throughout the section analyzed at Site 465 (Widmark & Malmgren 1992a). The Danian infaunal component at Site 465 did not only compensate for the loss of infaunal individuals across the K–PgB, but even outnumbered the epifaunal component in the Danian where *B. trinitatensis* became the dominant species (Widmark & Malmgren 1992a). It is also interesting to note that *B. trinitatensis* seems to represent an example of southward migration from low to (southern) high latitudes during the time close to the K–PgB event simultaneously with the drastical decline of typical Maastrichtian infaunal taxa, of which some even became (globally or locally) extinct. At the low-latitude Site 465, *B. trinitatensis* was present already at the Campanian–Maastrichtian boundary (Widmark, unpublished data), at the mid-latitude Site 525 it emerged during the terminal Maastrichtian (about 100 ka below the K–PgB) and crossed the transition (Widmark & Malmgren 1992a), and at the high-latitude Site 690 it was recorded in the basal Danian (Thomas 1990), showing a steady increase upward the secquence at all three localities.

What is behind this contradictory response to the K–PgB event by some infaunal taxa is hard to assess. It is obvious that some environmental factor(s) changed and that this change became devastating for some infauna but favorable for other infaunal taxa. Some of the favored species (such as *B. trinitatensis*, *B. velascoensis*, and *Aragonia* spp.) are characterized by very dense, heavily calcified test walls, which are quite different from the smooth and much less dense test walls of *Praebulimina*. One ecologic factor that may favor species with such heavily calcified tests could be the availability of certain elements such as Ca^+ in the ambient sea water. Caldeira & Rampino (1993) modelled the effects resulting from the mass mortality among (calcareous) plankton at the K–PgB, and they analyzed the ecologic aftermath of this catastrophic event. They suggested that the extinction of various calcareous plankton during the earliest Tertiary left an excess of Ca^+ (among other elements) in the sea water because the heavily reduced calcareous planktic biomass was unable to incorporate Ca^+ to the same degree as before the K–PgB event. It is, therefore, possible (although quite speculative) to suggest that such an excess of Ca^+ would favor blooms of heavily calcified species as *B. trinitatensis, B. velascoensis,* and *Aragonia* spp., in combination with an ability to profit from alternative food resources not utilizable by, for instance, *Praebulimina* and *Eouvigerina subsculptura,* and from the diminishing competition from typical Cretaceous infaunal taxa that became severely affected and reduced by the K–PgB event.

Summary and conclusion

The global or local extinctions among uppermost Cretaceous species (taxa) across the K–PgB range between 9% and 43%, implying that, in spite of the difficulties of comparing the various data sets reviewed herein, the foraminiferal benthos was much less affected by the biotic crisis at the boundary, than, e.g., the calcareous plankton such as coccoliths and planktic foraminifera, which lived closer to the surface.

The degree of extinctions seems to follow both paleobathymetric and paleolatitudinal gradients. For instance, the turnover rate in the middle- to outer-neritic environment at Brazos River was shown to be higher than those at upper-bathyal–outer-neritic environments within the Tethyan region (Keller 1992); however, this high turnover rate may be due to the controversial placement of the K–PgB made in the study (see above). In the deep-sea mid-latitude environment (Walvis Ridge, South Atlantic), Widmark & Malmgren (1992a) demonstrated that the degree of extinctions (taxonomic disappearances) was about twice as high at shallower, middle bathyal compared to deeper, upper abyssal depths; Thomas (1990) similarly recorded a higher extinction at shallower, middle bathyal than at deeper, lower bathyal depths in the high latitudes (Maud Rise, Southern Ocean). The paleolatitudinal response in terms of benthic foraminiferal extinctions is harder to assess, since difficulties arise when comparing quite shallow-water faunal data from the Tethyan marginal seas (Keller 1992; Speijer & Van der Zwaan 1996) with much deeper faunas from the Southern Ocean (Thomas 1990; Nomura 1991) because of the bathymetric component involved. Still we are able to compare the recalculated extinction (taxonomic disappearance) from the low-latitude middle-bathyal Site 465 with extinction data from the high-latitude Sites 689, 690, and 738, which reveals that low-latitude extinctions (22–

42%) are 2–3 times higher than in the high latitudes (9–14%; 12.3%). This suggests that tropical (low-latitude) faunas were more severely affected by the K–PgB event, but still more low-latitude bathyal data are needed to consistently verify this pattern.

A decline in primary productivity (i.e. cessation of the organic-carbon flux to the deeper environments) at the K–PgB has generally been accepted as an explanation for the benthic foraminiferal changes as indicated by the morphogroup analyses reviewed above, and in which certain dominating infaunal components, which indicate high food supply, drastically declined or even became extinct (globally or locally) at the K–PgB. Also the well-known mass extinctions among calcareous plankton (especially coccoliths and planktic foraminifera) at the K–PgB point toward a major biotic crisis at or near the surface. In asserting food to be the prime ecological factor controlling benthic foraminiferal distributions, we may explain why shallower bathyal faunas were more affected by the K–PgB event than deeper-bathyal to abyssal faunas – the former were simply better adapted to (and thus dependent upon) eutrophic conditions than the deeper, more or less oligotrophic communities and hence more vulnerable when the food supply failed at the K–PgB. Also, the lower degree of extinctions in the high latitudes could be explained by this in that high-latitude faunas are more generalistic (*K*-selected) in character with a smaller opportunistic (*r*-selected) component because of their distance to the (sub)tropical upwelling areas (Widmark 1995). With a smaller component of eutrophic, food-sensitive taxa these faunas will be less prone to a biotic crisis in terms of cessation in food supply and thus only have a minor effect on the fauna as a whole.

Finally, it has been argued, as mentioned before, that most deep-sea K–PgB sections are incomplete because of carbonate starvation (MacLeod & Keller 1991a, b) and that this should yield artificial high and abrupt faunal turnover and high extinctions in planktic foraminifera and coccoliths at most deep-sea cores compared to land-based sections (e.g., El Kef, Tunisia). If this is so, also extinctions among deep-sea benthic foraminifera should be higher than at, e.g., El Kef, which certainly is not the case. The recalculated extinction values from Site 465 (22–42%) are indeed compatible with an 'abrupt faunal turnover' involving an (at least local) extinction of 39% reported by Speijer & Van der Zwaan (1996) or even lower if the minimum taxonomic disappearance at Site 465 (22%) is taken into account. The minimum figure from Site 465 is actually the more proper figure to use in this context, since the El Kef figure is based on the exclusion of lumped taxa on the suprageneric level and thus closer to the minimum than to the maximum figure of taxonomic disappearance; the lower figure from Site 465 is also more compatible with the idea that deeper, oligotrophic or mesotrophic faunas would be less affected by food-supply failure than shallower, more eutrophic ones.

Systematic paleontology

Order Foraminiferida Eichwald, 1830

Diagnosis. – See Loeblich & Tappan (1988, p. 7).

Suborder Textulariina Delage & Hérouard, 1896

Diagnosis. – See Loeblich & Tappan (1988, p. 19).

Superfamily Astrorhizacea Brady, 1881 – Superfamily Hippocrepinacea Rhumbler, 1895

Diagnosis. – See Loeblich & Tappan (1988, p. 19 and 42, respectively).

Genera *Hyperammina* Brady, 1878 – *Bathysiphon* M. Sars *in* G.O. Sars, 1872

Type species and diagnosis. – See Loeblich & Tappan (1988, p. 42 and 22, respectively).

Hyperammina – Bathysiphon spp. (fragments)

Fig. 5A–B

Material. – About 150 fragments.

Occurrence. – Danian and Maastrichtian at Sites 525 and 527.

Remarks. – A relatively large number of fragmented agglutinated tubes encountered was assigned to this taxon.

Superfamily Ammodiscacea Reuss, 1862

Diagnosis. – See Loeblich & Tappan (1988, p. 46).

Family Ammodiscidae Reuss, 1862

Diagnosis. – See Loeblich & Tappan (1988, p. 46).

Subfamily Ammodiscinae Reuss, 1862

Diagnosis. – See Loeblich & Tappan (1988, p. 47).

Genus *Ammodiscus* Reuss, 1862

Type species and diagnosis. – See Loeblich & Tappan (1988, p. 47).

Ammodiscus cretaceus (Reuss, 1845)

Fig. 5C

Synonymy. – □*1845 *Operculina cretacea* Reuss, p. 35, Pl. 13:64–65. □1977 *Ammodiscus cretaceous* (Reuss) – Sliter, p. 677, Pl. 1:3. □1983 *Ammodiscus cretaceus* (Reuss) – Hart, p. 264, Pl. 1:4. □1988 *Ammodiscus cretaceus* (Reuss) – Kaminski *et al.*, p. 184, Pl. 3:7. □1990 *Ammodiscus cretaceus* (Reuss) – Charnock & Jones, p. 154, Pl. 2:1–2. □1990 *Ammodiscus cretaceus* (Reuss) – Klasz & Klasz, p. 404, Pl. 2:2. □1990 *Ammodiscus cretaceus* (Reuss) – Malata & Oszczypko, p. 515, Pl. 1:3. □1990 *Ammodiscus cretaceus* (Reuss) – Bellagamba & Coccioni, p. 899, Pl. 1:7. □1990 *Ammodiscus cretaceus* (Reuss) – Kuhnt, p. 310, Pl. 1:2–3. □1991 *Ammodiscus cretaceus* (Reuss) – Kuhnt & Moullade, p. 328, Pl. 4:A. □1992 *Ammodiscus cretaceus* (Reuss) – Kaiho, p. 242, Pl. 1:4. □1993 *Ammodiscus cretaceus* (Reuss) – Kuhnt & Kaminski, p. 72, Pl. 2:1.

Material. – Twenty-five specimens.

Description. – Test circular in outline. A single, undivided tubular chamber arranged in a planispiral coil, slowly increasing in diameter. Wall smooth and fine-grained. Aperture a simple opening at end of chamber.

Occurrence. – Danian and Maastrichtian at Sites 525 and 527.

Remarks. – Some rare specimens with discoidal test shapes and a tubular, undivided chamber arranged into a planispiral coil were included in this well-established species.

Subfamily Ammovertellininae Saidova, 1981

Diagnosis. – See Loeblich & Tappan (1988, p. 50).

Fig. 5. Scale 100 μm. □A–B. *Hyperammina–Bathysiphon* spp.; side views; sample 527-32-4, 57–59 cm (Maastrichtian). □A. Rectilinear specimen. □B. Irregular specimen with attached globotruncanid. □C. *Ammodiscus cretaceus* Reuss; dorsal view; sample 525A-40-2, 109–110 cm (Maastrichtian). □D–E. *Glomospira*? spp.; sample 527-32-4, 107–108 cm (Maastrichtian). □D. Ventral view of streptospiral specimen. □E. Dorsal view of irregular specimen.

Genus *Glomospira* Rzehak, 1885

Type species and diagnosis. – See Loeblich & Tappan (1988, p. 50).

Glomospira? spp.

Fig. 5D–E

Material. – Twenty-eight specimens.

Occurrence. – Maastrichtian at Site 525 and Danian and Maastrichtian at Site 527.

Remarks. – Some rare specimens with a tubular, undivided chamber arranged into an irregular or streptospiral coil were included in this taxon.

Superfamily Spiroplectamminacea Cushman, 1927

Diagnosis. – See Loeblich & Tappan (1988, p. 110).

Family Spiroplectamminidae Cushman, 1927

Diagnosis. – See Loeblich & Tappan (1988, p. 110).

Subfamily Spiroplectammininae Cushman, 1927

Diagnosis. – See Loeblich & Tappan (1988, p. 111).

Genus *Spiroplectammina* Cushman, 1927

Type species and diagnosis. – See Loeblich & Tappan (1988, p. 112).

Spiroplectammina dentata (Alth, 1850)

Fig. 6A

Synonymy. – □*1850 *Textularia dentata* sp. nov. – Alth, p. 262, Pl. 13:13. □1977 *Spiroplectammina dentata* (Alth) – Sliter, p. 675, Pl. 1:9. □1978 *Spiroplectammina dentata*

Fig. 6. Scale 100 µm. □A. *Spiroplectammina dentata* (Alth); side view; sample 527-32-4, 107–108 cm (Maastrichtian). □B. *Spiroplectammina israelskyi* Hillebrandt; side view, sample 525A-40-3, 70–71 cm (Maastrichtian). □C. *Spiroplectammina laevis* (Roemer); side view; sample 527-32-4, 107–108 cm (Maastrichtian). □D–E. *Spiroplectammina spectabilis* (Grzybowski); side views. □D. Large microsphere; sample 527-33-1, 18–20 cm (Maastrichtian). □E. Macrosphere; Sample 527-32-5, 33–34 cm (Maastrichtian).

(Alth) – Beckmann, p. 769, Pl. 1:4–5. □1983 *Spiroplectammina dentata* (Alth) – Dailey, p. 768, Pl. 1:1. □1988 *Spiroplectammina* sp. aff. *S. dentata* (Alth) – Kaminski *et al.*, p. 192, Pl. 7:10–11. □*v*1988 *Spiroplectammina dentata* (Alth) – Widmark & Malmgren, p. 65, Pl. 5:13. □1990 *Spiroplectammina dentata* (Alth) – Klasz & Klasz, p. 411, Pl. 5:6. □1990 *Spiroplectammina dentata* (Alth) – Kuhnt, p. 325, Pl. 6:14. □1991 *Spiroplectammina dentata* (Alth) – Kuhnt & Moullade, p. 324, Pl. 2:C. □1992 *Spiroplectammina dentata* (Alth) – Gawor-Biedowa, p. 27, Pl. 1:10. □*v*1992a *Spiroplectammina dentata* (Alth) – Widmark & Malmgren, p. 133, Pl. 10:4. □1993 *Spiroplectammina* ex gr. *dentata* (Alth) – Kuhnt & Kaminski, p. 75, Pl. 6:5–6.

Material. – Forty-four specimens.

Description. – Test 'kite'-shaped, initial end rounded, and rhomboid in section. Periphery angular, irregular, and 'dentate'. Early chambers planispirally arranged, later chambers are biserially arranged. Sutures hardly recognizable, oblique, and straight. Walls agglutinated and fairly coarse-grained. Aperture consisting of a simple, low arch at base of final chamber.

Occurrence. – Danian and Maastrichtian at Site 527.

Remarks. – Specimens having the typical spines at the border of the periphery are included in this species. Some specimens with the typical spines poorly preserved are also included herein.

Spiroplectammina israelskyi Hillebrandt, 1962

Fig. 6B

Synonymy. – □*1962 *Spiroplectammina israelskyi* n.sp. – Hillebrandt, p. 30–31, Pl. 1:5–7. □1990 *Spiroplectammina israelskyi* (Hillebrandt) – Kuhnt, p. 325, Pl. 6:16–17. □*v*1992a *Spiroplectammina* sp. B – Widmark & Malmgren, p. 133, Pl. 10:6. □1993 *Spiroplectammina israelskyi* (Hillebrandt) – Kuhnt & Kaminski, p. 75, Pl. 6:4.

Material. – Sixty-nine specimens.

Description. – Test 'kite'-shaped to elongated, initial end rounded, and rhomboid in section. Periphery angular with invaginations at sutures between chambers in biserial part of test. Early chambers planispirally coiled, later chambers biserially arranged. Sutures distinct and depressed in later biserial part of test; sutures oblique and straight. Walls agglutinated and fairly coarse-grained. Aperture consisting of a simple, low arch at the base of final chamber.

Occurrence. – Maastrichtian at Sites 525 and 527.

Remarks. – Under this species are specimens included that are more elongate than *S. dentata* and lack the peripheral spines of *S. dentata*. *Spiroplectammina israelskyi* differs from *S. laevis* in being less tapered and in having invaginations between the chambers along the periphery in the biserial part of the test.

Spiroplectammina laevis (Roemer, 1842)

Fig. 6C

Synonymy. – □*1842 *Textularia laevis,* n.sp. – Roemer, p. 97, Pl. 15:17. □1990 *Spiroplectammina* cf. *laevis* (Roemer) – Kuhnt, p. 325, Pl. 6:15. □1991 *Spiroplectammina laevis* (Roemer) – Kuhnt & Moullade, p. 324, Pl. 2:F–G. □v1992a *Spiroplectammina* sp. A – Widmark & Malmgren, p. 133, Pl. 10:5. □1993 *Spiroplectammina laevis* (Roemer) – Kuhnt & Kaminski, p. 75, Pl. 6:5.

Material. – Forty-four specimens.

Description. – Test 'kite'-shaped to elongated with a rounded initial end and rhomboid in section. Periphery angular and even. Early chambers arranged in a relatively small spiral, later chambers biserially arranged. Sutures distinct and depressed in later biserial part of test; sutures oblique and straight. Walls agglutinated, fairly coarse-grained. Aperture consisting of a simple, low arch at the base of final chamber.

Occurrence. – Maastrichtian at Site 525, Danian and Maastrichtian at Site 527.

Remarks. – This species is more elongate than *S. dentata* and lacks the peripheral spines of *S. dentata. Spiroplectammina laevis* differs from *S. israelskyi* in being more tapering and in having a smooth and even periphery.

Spiroplectammina spectabilis (Grzybowski, 1898)

Fig. 6D–E

Synonymy. – □*1898 *Spiroplecta spectabilis* sp. nov. – Grzybowski, p. 293, Pl. 1:6. □1978 *Bolivinopsis spectabilis* (Grzybowski) – Proto Decima & Bolli, p. 790, Pl. 1:3. □1983 *Spiroplectammina spectabilis* (Grzybowski) – Tjalsma & Lohmann, p. 20, Pl. 1:11; Pl. 9:8–10. □1983 *Spiroplectammina spectabilis* (Grzybowski) – Verdenius & Hinte, p. 195, Pl. 6:8, 11–12. □1984 *Spiroplectammina spectabilis* (Grzybowski) – Kaminski, p. 31, Pl. 12:1–9; Pl. 13:1–8 [with synonymy]. □1988 *Spiroplectammina spectabilis* (Grzybowski) – Jones, p. 148, Pl. 2:5. □1988 *Spiroplectammina spectabilis* (Grzybowski) – Kaminski *et al.,* p. 193, Pl. 7:16–18. □1990 *Spiroplectammina* aff. *spectabilis* (Grzybowski) – Kuhnt, p. 325, Pl. 6:18. □1990 *Spiroplectammina (Spiroplictinella) spectabilis* (Grzybowski) – Charnock & Jones, p. 182, Pl. 9:15–18. □1991 *Spiroplectammina spectabilis* (Grzybowski) – Nomura, p. 23, Pl. 1:26. □1992 *Spiroplectammina spectabilis* (Grzybowski) – Kaiho, p. 243, Pl. 1:12–13. □v1992a *Spiroplectammina spectabilis* (Grzybowski) – Widmark & Malmgren, p. 113, Pl. 10:7. □1993 *Spiroplectammina*

spectabilis (Grzybowski) – Kaminski & Geroch, p. 267, Pl. 12:4–7 [lectotype Pl. 12:4].

Material. – Seventy-seven specimens.

Description. – Test elongate, slender, and lenticulate in section. Periphery angular with faint keel. Early chambers in a planispiral coil, later chambers arranged in an elongate biserial stage. Sutures distinct and depressed, more or less straight in biserial part, curved backwards in coiled part. Walls agglutinated and mainly composed of fine-grained calcareous particles. Aperture consisting of a simple, low arch at the base of final chamber.

Occurrence. – Danian and Maastrichtian at Sites 525 and 527.

Remarks. – This species differs from other *Spiroplectammina* encountered in this material by its narrow and slender biserial stage with almost parallel sides. The morphologic variability of this species has been discussed by Kaminski (1984), who embraced more than ten similar species in this concept. A lectotype was designated by Kaminski & Geroch (1993).

Spiroplectammina spp. calcareous forms

Fig. 7A

Material. – Eleven specimens.

Occurrence. – Danian and Maastrichtian at Site 525 and Maastrichtian at Site 527.

Remarks. – Specimens with calcareous tests belonging to genus *Spiroplectammina* are included in this taxon.

Spiroplectammina spp. juvenile forms

Fig. 7B

Material. – Sixty specimens.

Occurrence. – Danian and Maastrichtian at Sites 525 and 527.

Remarks. – Some juvenile specimens of *Spiroplectammina* could not be consistently referred to species and were therefore included in this taxon.

Superfamily Verneuilinacea Cushman, 1911

Diagnosis. – See Loeblich & Tappan (1988, p. 129).

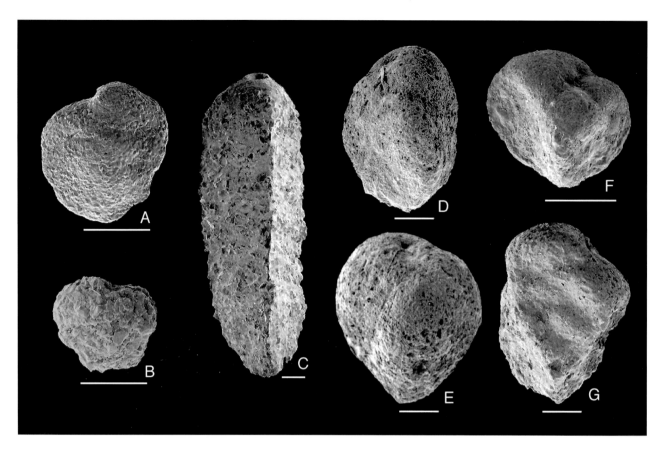

Fig. 7. Scale 100 µm. □A. *Spiroplectammina* spp. calcareous forms; side view; sample 525A-40-2, 59–60 cm (Maastrichtian). □B. *Spiroplectammina* spp. juvenile forms; side view; sample 525A-40-2, 100–102 cm (Maastrichtian). □C. *Tritaxia aspera* (Cushman); side view; sample 525A-40-2, 109–110 cm (Maastrichtian). □D–E. *Tritaxia* spp.; side views; sample 527-32-5, 92–94 cm (Maastrichtian). □F–G. *Gaudryina pyramidata* Cushman; sample 527-32-5, 33–34 cm (Maastrichtian). □F. Side view (juvenile specimen). □G. Side view (adult specimen).

Family Verneuilinidae Cushman, 1911

Diagnosis. – See Loeblich & Tappan (1988, p. 132).

Subfamily Verneuilininae Cushman, 1911

Diagnosis. – See Loeblich & Tappan (1988, p. 136).

Genus *Gaudryina* d'Orbigny

Type species and diagnosis. – See Loeblich & Tappan (1988, p. 136).

Gaudryina pyramidata Cushman, 1926

Fig. 7F–G

Synonymy. – □*1926b *Gaudryina laevigata* Franke var. *pyramidata* Cushman – Cushman, p. 587, Pl.16:8. □1977

Gaudryina pyramidata Cushman – Sliter, p. 675, Pl. 2:9. □1978 *Gaudryina pyramidata* Cushman – Beckmann, p. 766, Pl. 1:9–10. □1983 *Gaudryina pyramidata* Cushman – Dailey, p. 766, Pl. 1:6. □1983 *Gaudryina pyramidata* Cushman – Tjalsma & Lohmann, p. 12, Pl. 2:4; Pl. 8:1. □1988 *Gaudryina pyramidata* Cushman – Kaminski *et al.*, p. 194, Pl. 8:7. □*v*1988 *Gaudryina pyramidata* Cushman – Widmark & Malmgren, p. 65, Pl. 5:5. □1990 *Gaudryina pyramidata* Cushman – Kuhnt & Kaminski, p. 467, Pl. 5. figs. c–e, j. □1991 *Gaudryina pyramidata* Cushman – Kuhnt & Moullade, p. 326, Pl. 3:M–N. □1991 *Gaudryina pyramidata* Cushman – Nomura, p. 22, Pl. 5:11. □1992 *Gaudryina pyramidata* Cushman – Gawor-Biedowa, p. 34, Pl. 3:10. □*v*1992a *Gaudryina pyramidata* Cushman – Widmark & Malmgren, p. 111, Pl. 10:3. □*v*1992b *Gaudryina pyramidata* Cushman – Widmark & Malmgren, p. 393, Pl. 6:8. □1993 *Gaudryina pyramidata* Cushman – Kuhnt & Kaminski, p. 73, Pl. 6:8.

Material. – 226 specimens.

Description. – Test pyramidal and triangular–quadrate in section; periphery acute and undulating. Chambers visible and slightly inflated, triserial in early stage and biserial in later portion. Sutures distinct and slightly depressed. Walls agglutinated, relatively coarse-grained, but smooth. Aperture consisting of a low opening at the inner margin of last chamber.

Occurrence. – Danian and Maastrichtian at Site 525, Maastrichtian at Site 527.

Remarks. – This is a very stable form that was easy to recognize throughout the material because of its pyramidal test shape and triangular transection in the early stage of test. It is also a characteristic species of the 'Velasco-type' fauna.

Family Tritaxiidae Plotnikova, 1979

Diagnosis. – See Loeblich & Tappan (1988, p. 138).

Genus *Tritaxia* Reuss, 1860

Type species and diagnosis. – See Loeblich & Tappan (1988, p. 138).

Tritaxia aspera (Cushman), 1926

Fig. 7C

Synonymy. – □*v**1926b *Clavulina trilatera* Cushman var. *aspera* Cushman – Cushman, p. 589, Pl. 17:3. □1978 *Tritaxia insignis* (Cushman) – Beckmann, p. 769, Pl. 1:12. □1983 *Tritaxia aspera* (Cushman) – Dailey, p. 768, Pl. 1:8. □1988 *Clavulinoides aspera* (Cushman) – Kaminski *et al.*, p. 194, Pl. 8:11–12. □1990 *Tritaxia aspera* Cushman – Klasz & Klasz, p. 413, Pl. 6:6. □*v*1992a *Tritaxia aspera* Cushman – Widmark & Malmgren, p. 113, Pl. 10:10. □*v*1992b *Tritaxia aspera* Cushman – Widmark & Malmgren, p. 402, Pl. 6:7.

Material. – Eighty-seven specimens.

Description. – Test elongate, large, and stout; triangular in section. Periphery acute and even. Chambers triserial in early stage; uniserial in later portion. Sutures fairly distinct, slightly curved, and depressed in later uniserial portion. Walls agglutinated and rather coarse-grained. Aperture consisting of a rounded opening located terminally at a slight projection.

Occurrence. – Maastrichtian at Site 525.

Remarks. – Specimens assigned to this species are conspecific with the examined holotype of *Clavulina trilatera* Cushman var. *aspera* Cushman, and show close resemblance to the specimen of *T. aspera* (Cushman) illustrated by Dailey (1983). This species is common in, and characteristic of, the Maastrichtian of Site 525, but it is known to range into the Paleocene at other localities.

Tritaxia spp.

Fig. 7D–E

Material. – 310 specimens.

Occurrence. – Danian and Maastrichtian at Sites 525 and 527.

Remarks. – Representatives of this taxon are mostly small and juvenile. It was not possible to assign consistently the specimens referable to this taxon to a distinct species. Examination of the holotypes of *Clavulina amorpha* Cushman and *C. trilatera* Cushman led to the conclusion that some specimens referred to this taxon are probably referable to these species.

Superfamily Textulariacea Ehrenberg, 1838

Diagnosis. – See Loeblich & Tappan (1988, p. 168).

Family Eggerellidae Cushman, 1937

Diagnosis. – See Loeblich & Tappan (1988, p. 168).

Subfamily Dorothiinae Balakhmatova, 1972

Diagnosis. – See Loeblich & Tappan (1988, p. 168).

Genus *Dorothia* Plummer, 1931

Type species and diagnosis. – See Loeblich & Tappan (1988, p. 169).

Dorothia bulletta (Carsey, 1926)

Fig. 8A

Synonymy. – □*1926 *Gaudryina bulletta* sp. nov. – Carsey, p. 28, Pl. 4:4. □1946 *Dorothia bulletta* (Carsey) – Cushman, p. 46, Pl. 12:21–26. □1977 *Dorothia bulletta* (Carsey) – Sliter, p. 674, Pl. 3:7. □1978 *Dorothia bulletta* (Carsey) – Beckmann, p. 765, Pl. 1:18. □1983 *Dorothia bulletta* (Carsey) – Govindan & Sastri, p. 42, Pl. 4:1. □1990 *Dorothia bulletta* (Carsey) – Klasz & Klasz, p. 415, Pl. 7:1. □1990 *Dorothia bulletta* (Carsey) – Bellagamba &

Fig. 8. Scale 100 μm. □A. *Dorothia bulletta* (Carsey); side view; sample 525A-40-2, 109–110 cm (Maastrichtian). □B. *Dorothia pupa* (Reuss); side view; sample 525A-40-1, 130–131 cm (Danian). □C. *Marssonella oxycona* (Reuss); side view; sample 525A-40-2, 47–48 cm (Maastrichtian). □D. *Goesella rugosa* (Hanzlikova); side view; sample 527-32-4, 68–69 cm (Maastrichtian). □E. *Marssonella?* sp. triserial form; side view; sample 527-32-4, 87–89 cm (Maastrichtian).

Coccioni, p. 905, Pl. 2:20. □1991 *Dorothia bulletta* (Carsey) – Nomura, p. 22, Pl. 5:15a–b. □1992 *Dorothia bulletta* (Carsey) – Kaiho, p. 245, Pl. 1:14.

Material. – Twenty-five specimens.

Description. – Test elongate, almost cylindrical with a pointing initial end, subcircular–oval in section, and rounded periphery. Chambers of early stage trochospiral, later stage biserially arranged; apertural face (two last chambers) inflated. Sutures indistinct and slightly depressed. Walls agglutinated, consisting of fine-grained

sand or calcareous sand giving the test a dull, grey color. Aperture consisting of a simple arc at base of last chamber.

Occurrence. – Danian and Maastrichtian at Sites 525 and 527.

Remarks. – This form is separated from other *Dorothia* encountered in the present material by its inflated last chambers and later well-developed biserial stage, which gives this form its cylindrical test shape in lateral view.

Dorothia pupa (Reuss, 1860)

Fig. 8B

Synonymy. – □*1860 *Textularia pupa* sp. nov. – Reuss, p. 232, Pl. 13:4. □1978 *Dorothia pupa* (Reuss) – Beckmann, Pl. 765, Pl. 1:21. □1990 *Dorothia pupa* (Reuss) – Bellagamba & Coccioni, p. 906, Pl. 2:21–23. □1992 *Dorothia pupa* (Reuss) – Gawor-Biedowa, p. 56, Pl. 7:11. □*v*1992a *Dorothia pupa* (Reuss) – Widmark & Malmgren, p. 11, Pl. 10:1.

Material. – Sixty-one specimens.

Description. – Test stout and tapering, with pointing initial end; subcircular in transection with a broadly rounded periphery. Chambers in early portion trochospiral, later portion biserial. Chambers in biserial part strongly overlapping earlier parts of test. Sutures indistinct and hardly visible. Walls agglutinated, consisting of fine-grained sand or calcareous sand. Aperture consisting of a simple arc at base of final chamber.

Occurrence. – Danian and Maastrichtian at Site 525, Maastrichtian at Site 527.

Remarks. – *Dorothia pupa* is separated from other *Dorothia* by its stout test, subcircular outline, and strongly overlapping final chambers.

Genus *Marssonella* Cushman, 1933

Type species and diagnosis. – See Loeblich & Tappan (1988, p. 169).

Marssonella oxycona (Reuss, 1860)

Fig. 8C

Synonymy. – □*1860 *Gaudryina oxycona* sp. nov. – Reuss, p. 229, Pl. 12:3. □1946 *Marssonella oxycona* (Reuss) – Cushman, p. 43, Pl. 12:3–5. □1977 *Dorothia oxycona* (Reuss) – Sliter, p. 674, Pl. 3:8. □1978 *Dorothia* cf. *oxycona* (Reuss) – Beckmann, p. 765, Pl. 1:14–15. □1988 *Dorothia oxycona* (Reuss) – Govindan & Sastri, p. 43, Pl.

1:8. □1988 *Dorothia oxycona* (Reuss) – Kaminski *et al.*, p. 195, Pl. 9:9. □1990 *Marssonella oxycona* (Reuss) – Klasz & Klasz, p. 416, Pl. 7:7.

Material. – Fifty-seven specimens.

Description. – Test elongate, cylindrical to cone-shaped, with a pointing initial end; circular in section. Periphery rounded and even. Chambers in early stage trochospiral, later stage biserial; last two chambers (apertural face) flat, giving the test a truncated impression. Sutures indistinct and slightly depressed. Walls agglutinated, consisting of white to yellow fine-grained particles; walls smooth, especially at the flat apertural face. Aperture consisting of a simple arc at base of last chamber.

Occurrence. – Danian and Maastrichtian at Sites 525 and 527.

Remarks. – This species is easily separated from species of the similar genus *Dorothia* by its flat apertural face and its smooth test walls.

Marssonella? sp. triserial form

Fig. 8E

Material. – Forty-two specimens.

Description. – Test cone-shaped, with a pointing initial end; subtriangular in section and rounded periphery. Chambers in early part trochospiral; last whorl triserial, where the chambers are inflated toward the sides of apertural face. Sutures indistinct and hardly visible. Walls agglutinated, consisting of rather fine-grained particles. Aperture consisting of a simple arc at base of final chamber.

Occurrence. – Danian and Maastrichtian at Sites 525 and 527.

Remarks. – The taxonomic status of this form is uncertain. It shows some similarities with *D. oxycona*, and may represent the early (trochospiral to triserial) ontogeny of this species.

Family Valvulinidae Berthelin, 1880

Diagnosis. – See Loeblich & Tappan (1988, p. 181).

Subfamily Valvulininae Berthelin, 1880

Diagnosis. – See Loeblich & Tappan (1988, p. 182).

Genus *Goesella* Cushman, 1933

Type species and diagnosis. – See Loeblich & Tappan (1988, p. 183).

Goesella rugosa (Hanzlikova, 1953)

Fig. 8D

Synonymy. – □*1953 *Marssonella rugosa* n.sp. – Hanzlikova, p. 493, Pl. 2:5, 7. □1984 *Goesella rugosa* (Hanzlikova) – Geroch & Nowak, Pl. 4:8, 13, and 18. □v1988 *Dorothia trochoides* (Marsson) – Widmark & Malmgren, p. 67, Pl. 5:3. □1990 *Goesella rugosa* (Hanzlikova) – Kuhnt & Kaminski, p. 469, Pl. 5:k–m. □1991 *Goesella rugosa* (Hanzlikova) – Kuhnt & Moullade, p. 324, Pl. 2:L–K.

Material. – Forty-eight specimens.

Description. – Test cone-shaped, with rounded initial end; subcircular in section and rounded periphery. Chambers in early part trochospiral, the last two chambers biserially arranged. Sutures indistinct; only sutures of the last two (biserial) chambers visible, whereas sutures of earlier (trochospiral) chambers obscured by rather coarse-grained particles. Walls agglutinated, consisting of rather coarse-grained particles, except on apertural face, where the surface is smooth. Aperture consisting of an opening centrally at base of last chamber.

Occurrence. – Maastrichtian at Sites 525 and 527.

Remarks. – This species is separated from species of the similar genus *Dorothia* analyzed here by its large trochospiral portion and low number (usually two) of chambers in its later (biserial) part. *Goesella rugosa* differs from the morphologically similar *M. oxycona* by its stout test shape, coarse-grained test walls, rounded initial end, lower number of biserial chambers, and moderately inflated apertural face.

Suborder Spirillinina Hohenegger & Piller, 1975

Diagnosis. – See Loeblich & Tappan (1988, p. 303).

Family Patellinidae Rhumbler, 1906

Diagnosis. – See Loeblich & Tappan (1988, p. 305).

Subfamily Patellininae Rhumbler, 1906

Diagnosis. – See Loeblich & Tappan (1988, p. 306).

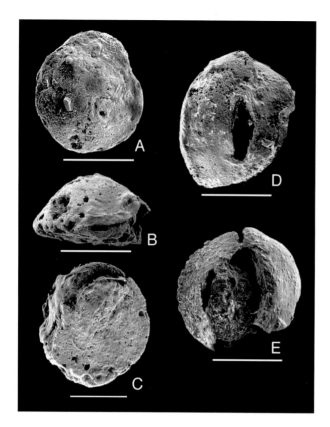

Fig. 9. Scale 100 μm. □A–C. *Patellina*? spp. □A. Spiral view; sample 525A-40-2, 69–70 cm (Maastrichtian). □B. Peripheral view; sample 525A-40-2, 69–70 cm (Maastrichtian). □C. Umbilical view; sample 525A-40-2, 78–79 cm (Maastrichtian). □D–E. 'Milioliniids'; sample 525A-40-2, 119–120 cm (Maastrichtian). □D. Dorsal view. □E. Ventral view.

Genus *Patellina* Williamson, 1858

Type species and diagnosis. – See Loeblich & Tappan (1988, p. 306).

Patellina? spp.
Fig. 9A–C

Material. – Twelve specimens.

Occurrence. – Danian and Maastrichtian at Site 525.

Remarks. – Only a few specimens that may be referable to this taxon were encountered in the present material.

Suborder Miliolina Delage & Hérouard, 1896

Diagnosis. – See Loeblich & Tappan (1988, p. 309).

'Miliolinids'
Fig. 9D–E

Material. – Five specimens.

Occurrence. – Maastrichtian at Sites 525 and 527.

Remarks. – A few, poorly preserved miliolinid specimen was found in a few samples at both sites analyzed.

Suborder Lagenina Delage & Hérouard, 1896

Diagnosis. – See Loeblich & Tappan (1988, p. 386).

Superfamily Nodosariacea Ehrenberg, 1838

Diagnosis. – See Loeblich & Tappan (1988, p. 394).

Family Nodosariidae Ehrenberg, 1838

Diagnosis. – See Loeblich & Tappan (1988, p. 394).

Subfamily Nodosariinae Ehrenberg, 1838

Diagnosis. – See Loeblich & Tappan (1988, p. 394).

Genus *Laevidentalina* Loeblich & Tappan, 1986

Type species and diagnosis. – See Loeblich & Tappan (1988, p. 396).

Laevidentalina spp.
Fig. 10A–B

Material. – Fifty-one specimens.

Occurrence. – Maastrichtian at Site 525, Maastrichtian and Danian at Site 527.

Remarks. – Several morphotypes, with uniserial chamber arrangement, slightly curved tests, oblique sutures, smooth (unornamented) test walls, and radiate, terminal, and eccentric apertures were lumped together under this taxon in accordance with Loeblich & Tappan (1988), who restricted *Dentalina* to concern only such species that have a longitudinally costate surface. *Laevidentalina* spp.

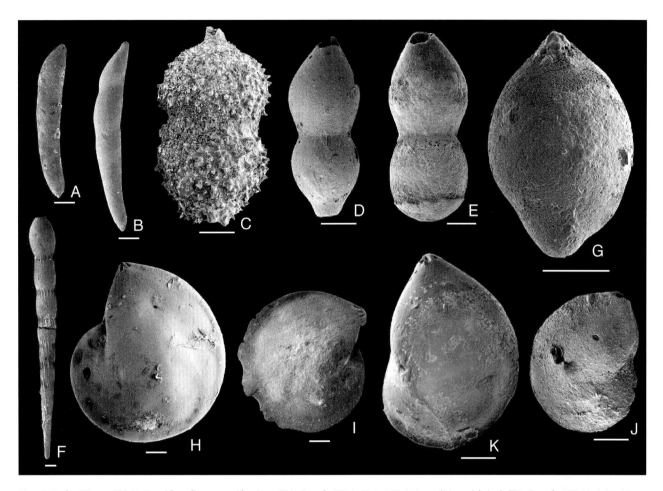

Fig. 10. Scale 100 µm. □A–B. *Laevidentalina* spp.; side views. □A. Sample 525A-40-2, 109–110 cm (Maastrichtian). □B. Sample 527-32-4, 87–89 cm (Maastrichtian). □C. '*Nodosaria*' *aspera* Reuss; side view; sample 527-32-4, 57–59 cm (Maastrichtian). □D–E. *Nodosaria* cf. *limbata* d'Orbigny; side views. □D. Specimen with initial spine preserved; sample 527-32-5, 57–58 cm (Maastrichtian). □E. Broken specimen; sample 525A-40-4, 60–61 cm (Maastrichtian). □F. '*Nodosaria*' *velascoensis* Cushman; side view; sample 525A-40-2, 109–110 cm (Maastrichtian). □G. *Pandaglandulina*? spp.; side view; sample 525A-40-2, 109–110 cm (Maastrichtian). □H. *Lenticulina* cf. *macrodisca* (Reuss); side view; sample 527-32-4, 57–59 cm (Maastrichtian). □I. *Lenticulina* cf. *rotulata* (Lamarck); side view; sample 525A-40-2, 109–110 cm (Maastrichtian). □J. *Lenticulina* spp.; side view; sample 525A-40-2, 47–48 cm (Maastrichtian). □K. *Saracenaria* spp.; side view; sample 525A-40-2, 78–79 cm (Maastrichtian).

occurs in relatively low abundances in most samples analyzed here.

Genus *Nodosaria* Lamarck, 1812

Type species and diagnosis. – See Loeblich & Tappan (1988, p. 397).

'*Nodosaria*' *aspera* Reuss, 1845

Fig. 10C

Synonymy. – □*1845 *Nodosaria aspera* nov. sp. – Reuss, p. 26, Pl. 13:14–15. □1932 *Nodosaria aspera* Reuss – Cushman & Jarvis, p. 35, Pl. 11:5. □1978 *Nodosaria aspera*

Reuss – Beckmann, p. 767, Pl. 1:24. □1992 *Nodosaria aspera* Reuss – Kaiho, p. 248, Pl. 2:4.

Material. – Two specimens.

Description. – Test elongate and circular in section. Periphery undulating owing to invaginations between chambers. Two to four globular, uniserially arranged chambers. Sutures orthogonal to axis of test, broad and distinct. Walls calcareous with hispid ornamentation. Aperture terminal and central.

Occurrence. – Maastrichtian at Site 527.

Remarks. – This species is easily recognized by its hispid appearance and chamber arrangement. A single specimen was encountered in two samples only. '*Nodosaria*' (with citation marks) are used to denote that the author is aware of the fact that Loeblich & Tappan (1988) restricted

Nodosaria to include species with smooth, unornamented tests only.

Nodosaria cf. *limbata* d'Orbigny, 1840

Fig. 10D–E

Synonymy. – □cf. *1840 *Nodosaria limbata* nov. sp. – d'Orbigny, p. 12, Pl. 1:1. □1932 *Nodosaria limbata* d'Orbigny – Cushman & Jarvis, p. 32, Pl. 10:5. □1978 *Nodosaria limbata* d'Orbigny – Beckmann, p. 767, Pl. 1:25. □1983 *Nodosaria limbata* d'Orbigny – Tjalsma & Lohmann, p. 16, Pl. 2:12.

Material. – Fifteen specimens.

Description. – Test elongate, slightly arcuate, circular in section. Periphery undulating owing to invaginations between chambers. Two to three globular, uniserially arranged chambers, the first one might have a short apical spine. Sutures distinct and perpendicular to long axis of test. Walls calcareous, smooth, and perforate. Aperture terminal, central, and radiate.

Occurrence. – Maastrichtian and Danian at Sites 525 and 527.

Remarks. – Close resemblance to *Nodosaria limbata* d'Orbigny given in the list of synonyms, but some specimens have an apical spine, which is absent in the figures illustrating this species in the papers of the synonyms given above.

'*Nodosaria*' *velascoensis* Cushman, 1926

Fig. 10F

Synonymy. – □*1926b *Nodosaria fontannesi* Berthelin, var. *velascoensis* Cushman – Cushman, p. 594, Pl. 18:12. □1932 *Nodosaria velascoensis* Cushman – Cushman & Jarvis, p. 35, Pl. 11:1–4. □1983 *Nodosaria velascoensis* Cushman – Dailey, p. 767, Pl. 2:1. □1983 *Nodosaria velascoensis* Cushman – Tjalsma & Lohmann, p. 16, Pl. 5:1.

Material. – Eleven specimens.

Description. – Test elongate, slightly arcuate, circular in section. Periphery undulating, owing to invaginations between chambers (obscured by ornamentations). Several uniserially arranged chambers, which are longer than broad; the earlier ones are heavily ornamented by costae. Sutures perpendicular to long axis of test; the costate ornamentation is restricted to the sutures in the later part of test. Walls calcareous, ornamentation consisting of fine longitudinal costae, which are somewhat spirally arranged. Aperture terminal and central on elongate neck.

Occurrence. – Maastrichtian and Danian at Site 525, Danian at Site 527.

Remarks. – This relatively rare species is easily recognized on the basis of its costate ornamentation. '*Nodosaria*' within quotation marks is used to indicate that the author is aware of the fact that Loeblich & Tappan (1988) restricted *Nodosaria* to include species with smooth, unornamented tests only.

Genus *Pandaglandulina* Loeblich & Tappan, 1955

Type species and diagnosis. – See Loeblich & Tappan (1988, p. 398).

Pandaglandulina? spp.

Fig. 10G

Material. – Three specimens.

Occurrence. – Maastrichtian at Site 525.

Remarks. – Three nodosariid specimens with robust test, strongly overlapping uniserial chambers, which are slightly arcuate in the initial end, and terminal, radiate apertures were included in this taxon.

Family Vaginulinidae Reuss, 1860

Diagnosis. – See Loeblich & Tappan (1988, p. 403).

Subfamily Lenticulininae Chapman, Parr, & Collins, 1934

Diagnosis. – See Loeblich & Tappan (1988, p. 404).

Genus *Lenticulina* Lamarck, 1804

Type species and diagnosis. – See Loeblich & Tappan (1988, p. 405).

Lenticulina cf. *macrodisca* (Reuss, 1862)

Fig. 10H

Synonymy. – □cf. 1862 *Cristellaria macrodisca* nov. sp. – Reuss, p. 78, Pl. 9:5. □v1932 *Robulus macrodiscus* (Reuss) – Cushman & Jarvis, p. 23, Pl. 7:3. □1983 *Lenticulina macrodisca* (Reuss) – Dailey, p. 767, Pl. 2:2.

Material. – Ninety-five specimens.

Description. – Test subcircular in outline and lenticular in section; involute and biumbonate with large umbos. Periphery angular and even. About six hardly visible chambers in the final whorl. Sutures hardly visible, slightly curved, and almost flush. Walls calcareous, rather opaque, and smooth. Aperture terminal and located at peripheral angle.

Occurrence. – Maastrichtian and Danian at Sites 525 and 527.

Remarks. – Specimens of *L.* cf. *macrodisca* in the samples analyzed here show a close resemblance to the specimen of *L. macrodisca* (Reuss) figured by Dailey (1983). *Lenticulina* cf. *macrodisca* is different from *L.* cf. *rotulata* (Lamarck) in lacking the sharp hyaline keel around the periphery, in being thicker, and in having larger and fewer chambers in the last whorl. The plesiotype given above and additional secondary types of *Robulus macrodiscus* (Reuss) (Cush. Coll. No:s 47962 and 61975) are well within the concept of the taxon recognized here.

Lenticulina cf. *rotulata* (Lamarck, 1804)

Fig. 10I

Synonymy. – □cf. *1804 ?*Lenticulites rotulata* n.sp. – Lamarck, p. 188, (1806), Pl. 62:11. □v1941 *Lenticulina rotulata* (Lamarck) – Cushman, p. 67, Pl. 16:13. □v1946 *Lenticulina rotulata* (Lamarck) – Cushman, p. 56, Pl. 19:5 and Pl. 19:3. □1943 *Lenticulina rotulata* (Lamarck) – Frizzell, p. 341, Pl. 54:2.

Material. – Ninety-four specimens.

Description. – Test subcircular in outline, lenticular in section, and involute. Periphery angular with sharp hyaline keel. Seven chambers in last whorl; chambers pointed at periphery and bent backwards. Sutures distinct, flush, weakly elevated, and curved. Walls calcareous and smooth with large portion of hyaline material between chambers; umbo of hyaline material through which chambers of inner whorl are visible. Aperture terminal and located at peripheral angle.

Occurrence. – Maastrichtian and Danian at Sites 525 and 527.

Remarks. – Specimens of *Lenticulina* with a sharp hyaline keel and a large hyaline boss are included in this taxon, and they show close resemblance to the specimen of *Lenticulina rotulata* (Lamarck) figured by Frizzell (1943). Among a large number of secondary types assigned to this species at the Cushman Collection, three plesiotypes were very close to the specimens referred to this taxon in having pointed chambers, which are bent backwards, flush to

weakly elevated sutures, and a well developed keel. The plesiotypes given above differ, however, from the specimens found here in being much larger in test size and in having a greater number of chambers in the last whorl (8–9, instead of about seven chambers).

Lenticulina spp.

Fig. 10J

Material. – Forty-four specimens.

Occurrence. – Maastrichtian and Danian at Sites 525 and 527.

Remarks. – A few specimens belonging to *Lenticulina* that were not enrolled throughout were included in this taxon.

Genus *Saracenaria* Defrance *in* de Blainville, 1824

Type species and diagnosis. – See Loeblich & Tappan (1988, p. 407).

Saracenaria spp.

Fig. 10K

Material. – Eleven specimens.

Occurrence. – Maastrichtian at Site 525.

Remarks. – Specimens with the characteristic saracenarian triangular transection and initially enrolled planispiral test were included in this taxon.

Subfamily Palmulinae Saidova, 1981

Diagnosis. – See Loeblich & Tappan (1988, p. 408).

Genus *Neoflabellina* Bartenstein, 1948

Type species and diagnosis. – See Loeblich & Tappan (1988, p. 409).

Neoflabellina spp.

Fig. 11A–B

Material. – Six specimens.

Occurrence. – Danian and Maastrichtian at Sites 525.

Remarks. – Only a few, mostly fragmented or juvenile specimens referable to the genus *Neoflabellina* were

Fig. 11. Scale 100 μm. □A–B. *Neoflabellina* sp.; side views; sample 525A-40-3, 70–71 cm (Maastrichtian). □A. Adult specimen. □B. Juvenile specimen. □C. *Citharina*? sp.; side view; sample 525A-40-4, 60–61 cm (Maastrichtian). □D. *Astacolus* spp., side view; sample 525A-40-2, 92–93 cm (Maastrichtian). □E. *Marginulina*? spp.; side view; sample 525A-40-1, 140–141 cm (Danian). □F. *Vaginulina*? sp.; side view; sample 527-32-4, 57–59 cm (Maastrichtian). □G. *Lagena* sp. ovate form; side view; sample 527-32-4, 29–30 cm (Danian). □H. *Lagena* cf. *sulcata* (Walker & Jacob); side view; sample 525A-40-2, 59–60 cm (Maastrichtian). □I. *Lagena* sp. globose form; side view; sample 527-32-4, 29–30 cm (Danian). □J. *Lagena* sp. hispid form; side view; sample 525A-40-2, 128–129 cm (Maastrichtian). □K. *Reussoolina apiculata* (Reuss); side view; sample 525A-40-2, 92–93 cm (Maastrichtian).

encountered in the material. At least two or three species, such as *N. buticula* Hiltermann, *N. permutata* Koch, and/ or *N. praereticulata* Hiltermann, are present in the samples analyzed, but the material is too poorly preserved to allow a proper identification of these species.

Subfamily Marginulininae Wedekind, 1937

Diagnosis. – See Loeblich & Tappan (1988, p. 409).

Genus *Astacolus* de Montfort, 1808

Type species and diagnosis. – See Loeblich & Tappan (1988, p. 410).

Astacolus spp.

Fig. 11D

Material. – Fifty-nine specimens.

Occurrence. – Maastrichtian and Danian at Sites 525 and 527.

Remarks. – A relatively low number of specimens with arcuate, sometimes partly enrolled, elongate tests, and radiate, terminal apertures are included in this taxon.

Genus *Marginulina* d'Orbigny, 1826

Type species and diagnosis. – See Loeblich & Tappan (1988, p. 411).

Marginulina? spp.

Fig. 11E

Material. – Two specimens.

Occurrence. – Danian at Site 525.

Remarks. – Two small, poorly preserved nodosariid specimens with early portion not completely enrolled and later portion rectilinear, uniserial chamber arrangement and oblique sutures, are included in this taxon.

Subfamily Vaginulininae Reuss, 1860

Diagnosis. – See Loeblich & Tappan (1988, p. 412).

Genus *Citharina* d'Orbigny *in* de la Sagra, 1839

Type species and diagnosis. – See Loeblich & Tappan (1988, p. 412).

Citharina? sp.

Fig. 11A–B

Material. – One single, fragmented specimen.

Occurrence. – Maastrichtian at Sites 525.

Remarks. – A single fragmented elongate specimen with smooth test walls was encountered in the present material and assigned to this taxon.

Genus *Vaginulina* d'Orbigny, 1826

Type species and diagnosis. – See Loeblich & Tappan (1988, p. 414).

Vaginulina? sp.

Fig. 11F

Material. – Six specimens.

Occurrence. – Maastrichtian and Danian at Site 527.

Remarks. – A few specimens with uniserial, arcuate tests, and radiate apertures located at dorsal angle, were allocated to this taxon.

Family Lagenidae Reuss, 1862

Diagnosis. – See Loeblich & Tappan (1988, p. 415).

Genus *Lagena* Walker & Jacob, 1798

Type species and diagnosis. – See Loeblich & Tappan (1988, p. 415).

Lagena cf. *sulcata* (Walker & Jacob), 1798

Fig. 11H

Synonymy. – □cf. *1798 *Serpula (Lagena) sulcata* nov. sp. – Walker & Jacob, p. 634, Pl. 14:5. □cf. 1964 *Lagena sulcata* (Walker & Jacob) – Loeblich & Tappan 1964, p. 518, Fig. 404:11).

Material. – Thirty-one specimens.

Description. – Test subglobular–ovate with processes pointing in opposite directions at apertural and apical ends; circular in section. Unilocular. Walls calcareous, ornamented by numerous costae. Aperture simple, terminal on elongate neck.

Occurrence. – Maastrichtian and Danian at Sites 525 and 527.

Remarks. – *Lagena sulcata* (Walker & Jacob) and its relatives are described in a number of varieties, subspecies, and species. Hermelin & Malmgren (1980) have shown by multivariate analysis that forms similar to *L. sulcata* probably belong to the same species and that the variety of forms is probably environmentally controlled. Therefore, no identification at the subspecific level was made for these forms.

Lagena sp. globose form

Fig. 11I

Material. – Fifty-four specimens.

Description. – Test subglobular–ovoid with rounded apical end, circular in section. Unilocular. Walls calcareous, smooth, and perforate (without ornamentation). Aperture simple, terminal on elongate neck.

Occurrence. – Maastrichtian and Danian at Sites 525 and 527.

Remarks. – Lagenid unilocular specimens with globular tests, which are close to the figured specimen named *L. globosa* Montague in Brotzen (1936), are included in this taxon.

Lagena sp. hispid form

Fig. 11J

Material. – Six specimens.

Description. – Test subglobular–ovate with processes pointing in opposite directions at apertural and apical ends; circular in section. Unilocular. Walls calcareous, ornamented by numerous spines. Aperture simple, terminal on elongate neck.

Occurrence. – Maastrichtian and Danian at Site 525.

Remarks. – Specimens with unilocular tests and hispid ornamentation are included in this species. *Lagena* sp. hispid form may represent the early, unilocular stage of '*Nodosaria*' *aspera*.

Lagena sp. ovate form

Fig. 11G

Material. – Forty-four specimens.

Description. – Test subglobular–ovate with processes pointing in opposite directions at apertural and apical ends; circular in section. Unilocular. Walls calcareous and smooth. Aperture simple, terminal on elongate neck.

Occurrence. – Maastrichtian and Danian at Sites 525 and 527.

Remarks. – Specimens with ovate to elongate tests and apical spines are referred to this taxon. *Lagena* sp. ovate form is distinguished from *Reussoolina apiculata* (Reuss) in having a simple aperture instead of the radiate aperture in *R. apiculata* (Reuss).

Genus *Reussoolina* Colom, 1956

Type species and diagnosis. – See Loeblich & Tappan (1988, p. 416).

Reussoolina apiculata (Reuss, 1851)

Fig. 11K

Synonymy. – □*1851b *Oolina apiculata* nov. sp. – Reuss, p. 22, Pl. 2:1. □1928a *Oolina apiculata* Reuss – White, p. 209, Pl. 29:6. □1936 *Lagena apiculata* (Reuss) – Brotzen, p. 109, Pl. 7:2. □1956 *Lagena apiculata* (Reuss) – Said & Kenawy, p. 136, Pl. 3:8. □1956 *Lagena reussi* Said & Kenawy, p. 145, Pl. 7:15.

Material. – One specimen.

Description. – Test subglobular to ovoid with short apical spine; circular in section. Unilocular. Walls calcareous, smooth, and perforate. Aperture terminal, central, and radiate.

Occurrence. – Maastrichtian at Site 525.

Remarks. – Some confusion has existed concerning the generic status of Reuss' *O. apiculata* as reflected by the list of synonyms above. *Oolina* has an entosolenian tube, whereas *Lagena* lacks a tube; *Oolina* may have a radiate aperture, whereas *Lagena* never has such an aperture (Loeblich & Tappan 1988). In order to solve the problem of placing Reuss' species, which lacks an entosolenian tube but has a radiate aperture, Loeblich & Tappan (1988) resurrected the subgenus of *Reussoolina* Colom (1926) and elevated it to be the generic status of *apiculata* (Reuss).

Family Polymorphinidae d'Orbigny, 1839

Diagnosis. – See Loeblich & Tappan (1988, p. 416).

Subfamily Falsoguttulininae Loeblich & Tappan, 1986

Diagnosis. – See Loeblich & Tappan (1988, p. 416).

Genus *Tobolina* Dain *in* Bykova *et al.*, 1958

Type species and diagnosis. – See Loeblich & Tappan (1988, p. 418).

Tobolia? spp.

Fig. 12A

Material. – Seventeen specimens.

Fig. 12. Scale 100 μm. □A. *Tobolia*? spp.; side view; sample 527-32-5, 92–94 cm (Maastrichtian). □B. *Globulina* cf. *lacrima* (Reuss); side view; sample 527-32-5, 92–94 cm (Maastrichtian). □C. *Globulina horrida* (Reuss); side view; sample 525A-40-2, 100–102 cm (Maastrichtian). □D–E. *Guttulina* spp.; dorsal views. □D. Slender specimen; sample 527-32-5, 92–94 cm (Maastrichtian). □E. Ovoid specimen; sample 527-32-4, 107–108 cm (Maastrichtian). □F. *Pyrolina*? spp.; side view; sample 527-32-4, 57–59 cm (Maastrichtian). □G. *Pyrolinoides*? spp.; side view; sample 527-32-4, 107–108 cm (Maastrichtian). □H. *Fissurina* spp.; side view; sample 527-32-4, 57–59 cm (Maastrichtian).

Occurrence. – Maastrichtian and Danian at Site 527.

Remarks. – The generic status of this taxon is uncertain. The specimens fit fairly well into the generic description of *Tobolia* given by Loeblich & Tappan (1988). The specimens encountered in the samples analyzed show a test shape intermediate between the test shapes of the specimens of *Globulina lacrima* and some ovoid specimens of *Guttulina* spp. encountered in the material.

Subfamily Polymorphininae d'Orbigny, 1839

Diagnosis. – See Loeblich & Tappan (1988, p. 418).

Genus *Globulina* d'Orbigny, 1839

Type species and diagnosis. – See Loeblich & Tappan (1988, p. 419).

Globulina cf. *lacrima* (Reuss, 1845)

Fig. 12B

Synonymy. – □cf. *1845 *Polymorphina (Globulina) lacrima* nov. sp. – Reuss, p. 40, Pl. 12:6; Pl. 13:83. □cf. 1943 *Globulina lacrima* (Reuss) – Frizzell, p. 348, Pl. 56:27. □cf. 1956 *Globulina lacrima* (Reuss) – Said & Kenawy, p. 137, Pl. 3:17–18. □*v*1988 *Globulina lacrima* (Reuss) – Widmark & Malmgren, p. 67, Pl. 1:2.

Material. – Seventy-one specimens.

Description. – Test ovate and circular in section. Three hardly visible chambers. Sutures obscured, but visible when moistened with water. Walls calcareous and smooth. Aperture small and difficult to recognize.

Occurrence. – Maastrichtian and Danian of Sites 525 and 527.

Remarks. – This species is distinguished from other unilocular forms by its rarely visible, but characteristic sutures.

Globulina horrida Reuss, 1846

Fig. 12C

Synonymy. – □*1846 *Globulina horrida* nov. sp. – Reuss, p. 100, Pl. 2:2. □1978 *Globulina lacrima horrida* (Reuss) – Beckmann, p. 766, Pl. 1:34. □1984 *Globulina lacrima horrida* (Reuss) – Nyong & Olsson, p. 473, Pl. 2:10.

Material. – Twenty-three specimens.

Description. – Test divided into two parts – early part ovate, later part smaller and having 4–5 tiny spines; circular in section. Three hardly visible chambers. Sutures obscured and hardly visible (as in *G.* cf. *lacrima*). Walls calcareous, smooth in early part, in later part rough. Aperture not observed in the few specimens encountered, or obscured by fistulous growth.

Occurrence. – Maastrichtian at Site 525, Maastrichtian and Danian at Site 527.

Remarks. – This distinct species is easy to recognize on the basis of its curious test shape.

Genus *Guttulina* d'Orbigny, 1839

Type species and diagnosis. – See Loeblich & Tappan (1988, p. 419).

Guttulina spp.

Fig. 12D–E

Material. – Thirty-nine specimens.

Occurrence. – Maastrichtian and Danian at Sites 525 and 527.

Remarks. – Several morphotypes of *Guttulina* are included in this taxon.

Genus *Pyrolina* d'Orbigny *in* de la Sagra, 1839

Type species and diagnosis. – See Loeblich & Tappan (1988, p. 421).

Pyrolina? spp.

Fig. 12F

Material. – Seventeen specimens.

Occurrence. – Maastrichtian at Site 525, Maastrichtian and Danian at Site 527.

Remarks. – Some less well preserved specimens were assigned to this taxon. *Pyrolina*? spp. is distinguished from *Pyrolinoides*? spp. in having a rounded rather than pointed initial end.

Genus *Pyrolinoides* Marie, 1941

Type species and diagnosis. – See Loeblich & Tappan (1988, p. 421).

Pyrolinoides? spp.

Fig. 12G

Material. – Eight specimens.

Occurrence. – Maastrichtian and Danian at Site 527.

Remarks. – Some less well preserved specimens were assigned to this taxon. *Pyrolinoides*? spp. is distinguished from *Pyrolina*? spp. in having a pointed rather than rounded initial end.

Family Ellipsolagenidae A. Silvestri, 1923

Diagnosis. – See Loeblich & Tappan (1988, p. 425).

Subfamily Ellipsolageninae A. Silvestri, 1923

Diagnosis. – See Loeblich & Tappan (1988, p. 428).

Genus *Fissurina* Reuss, 1850

Type species and diagnosis. – See Loeblich & Tappan (1988, p. 428).

Fissurina spp.

Fig. 12H

Material. – Twenty-five specimens.

Occurrence. – Maastrichtian at Site 525, Maastrichtian and Danian at Site 527.

Remarks. – A few rounded and compressed unilocular specimens with oval, slit-like apertures are included in this taxon. Some of the specimens referred to this taxon might be conspecific with *Fissurina orbignyana* Sequenza.

Suborder Rotaliina Delage & Hérouard, 1896

Diagnosis. – See Loeblich & Tappan (1988, p. 496).

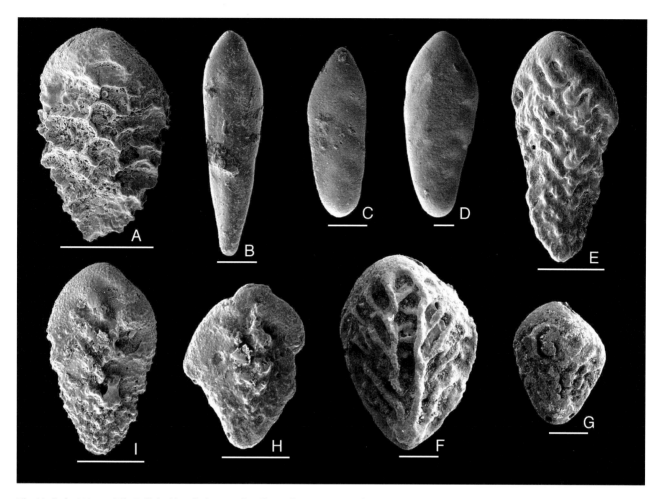

Fig. 13. Scale 100 μm. □A. *Bolivinoides* cf. *clavatus* Plotnikova; front view; sample 527-32-5, 33–34 cm (Maastrichtian). □B–D. *Brizalina incrassata* (Reuss); front vierws. □B. Specimen referable to the slender '*limonense*'-morphotype; sample 525A-40-3, 70–71 cm (Maastrichtian). □C. Specimen referable to the normal '*incrassata*'-morphotype; sample 525A-40-2, 100–102 cm (Maastrichtian). □D. Specimen referable to the large '*crassa*'-morphotype; sample 525A-40-4, 60–61 cm (Maastrichtian). □E. *Bolivinoides delicatula* Cushman; front view; sample 525A-40-2, 69–70 cm (Maastrichtian). □F. *Bolivinoides draco* (Marsson); front view; sample 525A-40-3, 130–131 cm (Maastrichtian). □G. *Bolivinoides decoratus* (Jones); front view; sample 525A-40-3, 70–71 cm (Maastrichtian). □H. *Bolivinoides paleocenicus* (Brotzen); front view; sample 525A-40-1, 120–121 cm (Danian). □I. *Bolivinoides* cf. *postulatus* Reiss; front view; sample 527-32-4, 107–108 cm (Maastrichtian).

Superfamily Bolivinacea Glaesser, 1937

Diagnosis. – See Loeblich & Tappan (1988, p. 497).

Family Bolivinidae Glaesser, 1937

Diagnosis. – See Loeblich & Tappan (1988, p. 497).

Genus *Brizalina* Costa, 1856

Type species and diagnosis. – See Loeblich & Tappan (1988, p. 498).

Brizalina incrassata (Reuss, 1851)

Fig. 13B–D

Synonymy. – □*1851b *Bolivina incrassata* nov. sp. – Reuss, p. 45, Pl. 5:13. □1969 *Brizalina incrassata* (Reuss) – Fisher, p. 194. □1983 *Coryphostoma incrassata* (Reuss) – Dailey, p. 766, Pl. 4:4. □1986 *Coryphostoma incrassata* (Reuss) – Van Morkhoven *et al.*, p. 382, Pl. 126. □1969 *Coryphostoma incrassata* (Reuss) – Nomura, p. 22, Pl. 1:3. □1992 *Bolivina incrassata* (Reuss) – Gawor-Biedowa, p. 95, Pl. 19:1–3.

Material. – Fifty-one specimens.

Description. – Test elongate, tapering, and oval in section. Periphery smooth and slightly angular. Chambers biserially arranged. Sutures visible throughout; straight without crenulations. Walls calcareous, perforate, and smooth without ornamentation. Aperture loop-shaped at base of final chamber, extending upwards.

Occurrence. – Maastrichtian and Danian at Sites 525 and 527.

Remarks. – Fisher (1969) referred this species to the genus *Brizalina*, because it lacks both retral processes and sutural crenulations, which are typical characters of *Bolivina* (Loeblich & Tappan 1988). Dailey (1983) and some other workers referred this form to *Coryphostoma*, the generic characters of which are roundness in section and the tendency towards uniseriality in the later stage of test (Loeblich & Tappan 1988). These characters were not observed on the specimens examined here. Some specimens with very slender tests, which might be referred to *Bolivina incrassata* Reuss var. *limonense* Cushman, were also included in this species, as well as some large specimens that might be referred to as *Bolivina incrassata* Reuss var. *crassa* Vasilenko & Myatlyuk, and of which the former morph may represent the microspheric generation and the latter an ecofenotypical variety of *B. incrassata*.

Family Bolivinoididae Loeblich & Tappan, 1984

Diagnosis. – See Loeblich & Tappan (1988, p. 500).

Genus *Bolivinoides* Cushman, 1927

Type species and diagnosis. – See Loeblich & Tappan (1988, p. 500).

Bolivinoides cf. *clavatus* Plotnikova, 1967
Fig. 13A

Synonymy. – □cf. 1992 *Bolivinoides clavatus* Plotnikova – Gawor-Biedowa, p. 98, Pl. 19:10–11.

Material. – Twenty-nine specimens.

Description. – Test elongate, compressed, tapering, and oval in section. Periphery uneven and rough by ornamentation. Chambers biserially arranged, but hardly visible because of ornamentation. Sutures almost obscured by the crenulating ornamentation, which forms a reticulate pattern all over the test, except for the last two chambers. Walls calcareous, perforate, and ornamented. Aperture consisting of a simple slit extending from base of last chamber.

Occurrence. – Danian and Maastrichtian at Sites 525 and 527.

Remarks. – The specific status of this rare but quite stable form is somewhat uncertain. It resembles, however, the figured specimen of *Bolivinoides clavatus* in Gawor-Biedowa (1992) on the basis of the rough, reticulate ornamentation situated over most parts of the test.

Bolivinoides decoratus (Jones, 1886)
Fig. 13G

Synonymy. – □*1886 *Bolivina decorata* n.sp. – Jones, p. 330, Pl. 27:7–8. □1950 *Bolivinoides decorata decorata* (Jones) – Hiltermann & Koch, p. 606, Text-figs. 2–4, Nos. 14–25, 27–31, 35–38, 42–45; Text-fig. 5, Nos. 36, 71. □1954 *Bolivinoides decorata decorata* (Jones) – Reiss, p. 155, Pl. 28:5–8, 13. □1992 *Bolivinoides decoratus* (Jones) – Gawor-Biedowa, p. 99, Pl. 20:4. □v1994 *Bolivinoides decoratus* (Jones) – Speijer, p. 46, Pl. 9:2.

Material. – Twenty-three specimens.

Description. – Test rhomboidal, compressed, and lenticular in section. Periphery angular and uneven. Chambers biserially arranged; the numerous and narrow chambers partly obscured by ornamentation. Sutures obscured by ornamentation, but visible at the margin; sutures oblique and slightly curved. Walls calcareous and heavily ornamented by a number of knobs, which longitudinally are arranged into two rows. Aperture loop-shaped at base of final chamber, extending upwards partly across the apertural face.

Occurrence. – Maastrichtian at Site 525.

Remarks. – This species is easy to recognize and differs from *B. draco* by its two longitudinal rows of knobs, which instead are fused into two longitudinal costae in *B. draco*.

Bolivinoides delicatulus Cushman, 1927
Fig. 13E

Synonymy. – □*1926b *Bolivina decorata* Cushman (non Jones) Cushman, p. 586, Pl. 15:11. □v1927a *Bolivinoides decorata* (Jones) var. *delicatula* Cushman – Cushman, p. 90, Pl. 12:8. □1954 *Bolivinoides delicatula* Cushman – Reiss, p. 157, Pl. 31:4. □1983 *Bolivinoides delicatulus* Cushman – Dailey, p. 766, Pl. 2:5. □1986 *Bolivinoides delicatulus* Cushman – Van Morkhoven *et al.*, p. 337, Pl. 110. □1990 *Bolivinoides delicatulus* Cushman – Thomas, p. 589, Pl. 1:15. □1991 *Bolivinoides delicatulus* Cushman – Nomura, p. 21, Pl. 1:6.

Material. – Seventeen specimens.

Description. – Test rhomboid and oval in section, with a rounded and smooth periphery. Chambers biserially arranged and hardly visible because of ornamentation. Sutures oblique, broad, and obscured by sets of 3–4 lobe-like crenulations. Walls calcareous; the lobe-like, crenulated ornamentation has a more delicate appearance than in other *Bolivinoides*. Aperture consisting of a loop-shaped opening, extending upwards from base of last chamber.

Occurrence. – Maastrichtian and Danian at Site 525.

Remarks. – The specimens assigned to this species are very close to the holotype of *Bolivinoides decorata* (Jones) var. *delicatula* Cushman examined at the Cushman Collection. Both the holotype and the specimens encountered in the present material were found to be conspecific, and they all have 3–4 lobe-like crenulations at the base of each chamber. This species is distinguished from other *Bolivinoides* described here by its slender outline of test, oval transection, and delicate loop-like ornamentation.

Bolivinoides draco (Marsson, 1878)

Fig. 13F

Synonymy. – □*1878 *Bolivina draco* n.sp. – Marsson, p. 157, Pl. 3:25. □v*1926a *Bolivina rhomboidea* n.sp. – Cushman, p. 19, Pl. 12:10. □1950 *Bolivinoides draco draco* (Marsson) – Hiltermann & Koch, p. 598, Text-fig. 1, No. 72a–c, 73a–b; Text-figs. 2–4, No. 52–54, 58–60; Text-fig. 5, No. 53, 69, 70. □1954 *Bolivinoides draco draco* (Marsson) – Reiss, p. 155, Pl. 29:13. □1983 *Bolivinoides draco draco* (Marsson) – Dailey, p. 766, Pl. 2:8. □1986 *Bolivinoides draco draco* (Marsson) – Van Morkhoven *et al.*, p. 378, Pl. 124. □1990 *Bolivinoides draco* (Marsson) – Nomura, p. 21, Pl. 1:5. □1982 *Bolivinoides draco* (Marsson) – Gawor-Biedowa, p. 101, Pl. 20:8. □v1994 *Bolivinoides draco* (Marsson) – Speijer, p. 48, Pl. 1:7.

Material. – Seventy-two specimens.

Description. – Test rhomboidal, compressed, and lenticular in section. Periphery angular and uneven. Chambers biserially arranged; the numerous and narrow chambers partly obscured by ornamentation. Sutures obscured by ornamentation, but visible at the margin; sutures oblique and slightly curved. Walls calcareous and heavily ornamented by a number of knobs, which longitudinally are fused into two more or less continuous costae. Aperture consists of a loop-shaped opening at base of final chamber, extending upwards partly across the apertural face.

Occurrence. – Maastrichtian at Site 525.

Remarks. – This species is easy to recognize and differs from *B. decoratus* by its two longitudinal costae, which instead are divided into rows of knobs in *B. decoratus*. The

holotype of a similar 'species', *B. rhomboidea* Cushman, was examined at the Smithsonian Institution. I am not convinced that *B. draco* Marsson and *B. rhomboidea* Cushman are separate species, since the short test and the peripheral thickening, which are diagnostic of *B. rhomboidea* (Cushman), may be well within the limits of the intraspecific variation of *B. draco* Marsson. *Bolivinoides draco* Marsson should have priority in being the senior synonym.

Bolivinoides paleocenicus (Brotzen, 1948)

Fig. 13H

Synonymy. – □*1948 *Bolivina paleocenica* n.sp. – Brotzen, p. 66, Pl. 9:5. □1954 *Bolivinoides paleocenica* Brotzen – Reiss, p. 157, Pl. 30:9–11. □1986 *Aragonia* sp. 1 – Van Morkhoven *et al.*, p. 374, Pl. 122. □1992 *Bolivinoides paleocenicus* (Brotzen) – Gawor-Biedowa, p. 104, Pl. 20:3.

Material. – Six specimens.

Description. – Test rhomboid, compressed, and lenticular in section. Periphery acute and uneven. Chambers biserially arranged, narrow and numerous. Sutures distinct and obscured by ornamentation only at the central part of test. Walls calcareous; ornamentation consisting of 2–3 raised ribs at each suture. Aperture hardly visible at the few specimens encountered in the material.

Occurrence. – Danian at Site 525.

Remarks. – This rare species is distinguished from other *Bolivinoides* described here on the basis of its small test size, acute periphery, compressed lenticular transection, 'kite'-shaped outline in front view, and weak ornamentation. A large number of secondary types of *Bolivina paleocenica* Brotzen were available at Brotzen's collection at the Naturhistoriska Riksmuseet, Stockholm. The vast majority of these specimens were very close to the specimens here referred to this species.

Bolivinoides cf. *postulatus* Reiss, 1954

Fig. 13I

Synonymy. – □cf. 1992 *Bolivinoides postulatus* Reiss – Gawor-Biedowa, p. 106, Pl. 19:7–8.

Material. – Fourteen specimens.

Description. – Test rhomboid, compressed, and lenticular in section. Periphery angular and uneven. Chambers biserially arranged; only axial part of chambers obscured by ornamentation. Sutures distinct, except at the axial part of test, which is obscured by ornamentation. Walls calcareous; ornamentation consisting of one or two large loop-shaped crenulation per chamber, located along the axial part of test; initial part of test covered with small postules. Aperture loop-shaped and located at base of final chamber, extending upwards partly across the apertural face.

Occurrence. – Danian(?) and Maastrichtian(?) at Site 525, Maastrichtian at Site 527.

Remarks. – This is the only *Bolivinoides* that occurred at the deeper Site 527, and the two poorly preserved specimens found at the shallower Site 525 are somewhat uncertain taxonomically. The present form is comparable to the specimen of *B. postulatus* Reiss figured by Gawor-Biedowa (1992) in its over-all test shape and in having small postules, especially on the initial part of test.

Superfamily Loxostomatacea Loeblich & Tappan, 1962

Diagnosis. – See Loeblich & Tappan (1988, p. 500).

Family Loxostomatidae Loeblich & Tappan, 1962

Diagnosis. – See Loeblich & Tappan (1988, p. 500).

Genus *Aragonia* Finlay, 1939

Type species and diagnosis. – See Loeblich & Tappan (1988, p. 500).

Aragonia spp.
Fig. 14A–B

Material. – Fifty-eight specimens.

Occurrence. – Danian and Maastrichtian at Sites 525 and 527.

Remarks. – Species such as *A. ouezzaensis* (Rey), *A. velascoensis* (Cushman), and *A. trinitatensis* (Cushman & Jarvis) are included in this taxon. The different state of preservation of the specimens studied here made a consistent separation of these species impossible.

Genus *Loxostomum* Ehrenberg, 1854

Type species and diagnosis. – See Loeblich & Tappan (1988, p. 500).

Loxostomum sp.
Fig. 14C

Material. – Twenty-eight specimens.

Description. – Test elongate, compressed with flat to concave sides; quadrangular in section. Periphery angular with narrow hyaline boarder. Chambers biserially arranged throughout, with a tendency to uniserial arrangement in later portion. Sutures distinct, limbate, and arched. Walls calcareous and finely perforate. Aperture not observed in the few, poorly preserved specimens encountered in the material analyzed here.

Occurrence. – Maastrichtian at Site 525.

Remarks. – Only a few broken specimens of this taxon were found in the material studied, and they were compared to the holotypes of *Bolivinita eleyi* Cushman (Cush. Coll. No. 5552) and *B. planata* Cushman (Cush. Coll. No. 6701). *Bolivinita eleyi*, which was described in Cushman (1927a), seems to be thicker in side view and less tapering than the specimens found here. The other holotype compared, *Bolivinita planata* (described in Cushman 1927b), is closer to the specimens encountered in the present material with regard to the more tapering and compressed test.

Superfamily Eouvigerinacea Cushman, 1927

Diagnosis. – See Loeblich & Tappan (1988, p. 509).

Family Eouvigerinidae Cushman, 1927

Diagnosis. – See Loeblich & Tappan (1988, p. 510).

Genus *Eouvigerina* Cushman, 1926

Type species and diagnosis. – See Loeblich & Tappan (1988, p. 510).

Eouvigerina subsculptura McNeil & Caldwell, 1981
Fig. 14D

Synonymy. – □*v**1933 *Eouvigerina aculeata* n.sp. – Cushman, p. 62, Pl. 7:8. □1981 *Eouvigerina subsculptura* nom. nov. – McNeil & Caldwell, p. 231, Pl. 18:20–21. □1983 *Eouvigerina americana* Cushman – Dailey, p. 766, Pl. 2:11. □*v*1992a *Eouvigerina subsculptura* McNeil & Caldwell – Widmark & Malmgren, p. 111, Pl. 1:8. □*v*1992b *Eouvige-*

Fig. 14. Scale 100 µm. □A–B. *Aragonia* spp.; front views. □A. Specimen referable to the inflated *'ouezzanensis'*-morphotype; sample 527-32-4, 57–59 cm (Maastrichtian). □B. Specimen referable to the normal *'velascoensis'*-morphotype; sample 525A-40-2, 69–70 cm (Maastrichtian). □C. *Loxostomum* sp.; front view; sample 525A-40-2, 119–120 cm (Maastrichtian). □D. *Eouvigerina subsculptura* McNeil and Caldwell; side view; sample 525A-40-3, 130–131 cm (Maastrichtian). □E. *Praebulimina* spp.; side view; sample 525A-40-3, 130–131 cm (Maastrichtian). □F. *Praebulimina reussi* (Morrow); side view; sample 525A-40-3, 130–131 cm (Maastrichtian). □G. *Pseudouvigerina plummerae* Cushman; side view; sample 525A-40-4, 60–61 cm (Maastrichtian). □H. *Pyramidina* sp.; side view; sample 525A-40-1, 130–131 cm (Danian).

rina subsculptura McNeil & Caldwell – Widmark & Malmgren, p. 393, Pl. 1:8. □*v*1994 *Eouvigerina subsculptura* McNeil & Caldwell – Speijer, p. 48, Pl. 1:7.

Material. – 764 specimens.

Description. – Test elongate, tapering, compressed initially, and slightly twisted. Periphery rounded. Chambers biserially arranged and indistinct in early portion, in later portion distinct and with a tendency towards uniseriality; lower boarder of the later chambers marked by a concave rim. Sutures indistinct and depressed. Walls calcareous and ornamented by short tiny spines. Aperture terminal, located on a short neck.

Occurrence. – Maastrichtian at Site 525.

Remarks. – Dailey (1983) figured the same form under the name of *E. americana* Cushman. However, the holotype of *E. americana* Cushman (Cush. Coll. No. 4986) is much more sculptured and smooth than the specimens found in the present material. The holotype of a similar species, *E. aculeata* Cushman, was examined and found to be close to the specimens encountered here. McNeil & Caldwell (1981) renamed *E. aculeata* Cushman, since it was a junior homonym of *E. aculeata* (Ehrenberg), suggesting the name *E. subsculptura* (in order to point out the less sculptured appearance of this form). These authors also noted that in the figured holotype of *E. aculeata* Cushman, the initial (biserial) chambers were poorly defined. In their own material, they observed that this part of the test was covered with fine rugae, a feature, which was found in the present material also. *Eouvigerina subsculptura* is one of the most abundant species in the Maastrichtian samples of Site 525.

Superfamily Turrilinacea Cushman, 1927

Diagnosis. – See Loeblich & Tappan (1988, p. 511).

Family Turrilinidae Cushman, 1927

Diagnosis. – See Loeblich & Tappan (1988, p. 511).

Genus *Praebulimina* Hofker, 1953

Type species and diagnosis. – See Loeblich & Tappan (1988, p. 511).

Praebulimina reussi (Morrow, 1934)
Fig. 14F

Synonymy. – □*1934 *Bulimina reussi* n.sp. – Morrow, p. 195, Pl. 29:12. □*v*1935 *Bulimina reussi* Morrow – Cushman & Parker, p. 99, Pl. 15:8. □*v*1941 *Bulimina reussi* Morrow – Cushman & Hedberg, p. 95, Pl. 22:30. □*v*1944 *Bulimina reussi* Morrow – Cushman & Deaderick, p. 337, Pl. 53:6. □1983 *Praebulimina reussi* (Morrow) – Dailey, p. 768, Pl. 2:7. □1992 *Praebulimina reussi* (Morrow) – Gawor-Biedowa, p. 116, Pl. 21:5–8. □*v*1994 *Praebulimina reussi* (Morrow) – Speijer, p. 48, Pl. 1:11.

Material. – About 200 specimens.

Description. – Test fusiform and subcircular in section. Periphery rounded and smooth. Chambers triserially arranged, initial end rounded, thereafter rapidly flaring; chambers of last whorl inflated. Sutures distinct, weakly curved, and slightly depressed. Walls calcareous and perforate. Aperture subterminal, narrow, and comma-shaped; aperture extends from base of last chamber to periphery in front view.

Occurrence. – Maastrichtian at Sites 525 and 527.

Remarks. – Several secondary types identified as *Bulimina reussi* Morrow [= *Praebulimina reussi* (Morrow)] were examined at the Cushman Collection (Smithsonian Institution) and found to be consistent with the concept applied here for this species. This species was distinguished from other *Praebulimina* encountered on the basis of its apertural features (subterminal), since test shape is not a good distinguishing character at the specific level in these highly variable organisms.

Praebulimina spp.
Fig. 14E

Material. – About 170 specimens.

Occurrence. – Maastrichtian at Site 525, Maastrichtian and Danian at Site 527.

Remarks. – This taxon includes morphotypes of *Praebulimina* with fusiform test shapes, which are difficult to separate consistently, and which lack the typical subterminal aperture of *P. reussi* (Morrow). Some specimens with damaged apertural parts are also included.

Genus *Pseudouvigerina* Cushman, 1927

Type species and diagnosis. – See Loeblich & Tappan (1988, p. 511).

Pseudouvigerina plummerae Cushman, 1927
Fig. 14G

Synonymy. – □*v**1927b *Pseudouvigerina plummerae* n.sp. – Cushman, p. 115, Pl. 23:8. □1983 *Pseudouvigerina plummerae* Cushman – Dailey, p. 768, Pl. 3:1. □*v*1994 *Pseudouvigerina plummerae* Cushman – Speijer, p. 48, Pl.4:2.

Material. – Fifty-eight specimens.

Description. – Test elongate and triangular in section. Periphery undulated. Chambers triserially arranged; chambers slightly inflated, truncated at the periphery, and having crenulated margins. Sutures distinct and slightly depressed. Walls calcareous, smooth, and finely perforate. Aperture terminal and circular.

Occurrence. – Maastrichtian at Site 527.

Remarks. – The specimens allocated to this species are close to the holotype of *Pseudouvigerina plummerae* Cushman, and they were found to be consistent with the concept of this species used here. This species is easy to recognize on the basis of its triangular transection, truncated chamber margins, and terminal aperture on a short neck, which may be broken off in some specimens.

Genus *Pyramidina* Brotzen, 1948

Type species and diagnosis. – See Loeblich & Tappan (1988, p. 512).

Pyramidina sp.

Fig. 14H

Material. – Forty-two specimens.

Description. – Test tapering and subtriangular in section, with concave sides; test occasionally slightly twisted along its long axis. Periphery subangular and uneven. Chambers triserially arranged; low and broad. Sutures distinct, depressed, and sigmoid at base of chambers. Walls calcareous and finely perforate; wall surface somewhat nodose. Aperture loop-shaped, extending from base of last chamber.

Occurrence. – Maastrichtian and Danian at Site 525, Danian at Site 527.

Remarks. – Unfortunately, the holotype of a similar species, *Bulimina rudita* Cushman, was missing from the Cushman Collection. However, the paratypes of this species were available in the collection and found to be different from the specimens of *Pyramidina* sp. encountered in the present material. *Pyramidina* sp. is different from *P. rudita* in its more distinctly triangular transection, acute edges, and more pointing initial end.

Superfamily Buliminacea Jones, 1875

Diagnosis. – See Loeblich & Tappan (1988, p. 515).

Family Buliminidae Jones, 1875

Diagnosis. – See Loeblich & Tappan (1988, p. 521).

Genus *Bulimina* d'Orbigny, 1826

Type species and diagnosis. – See Loeblich & Tappan (1988, p. 521).

Bulimina spinea Cushman & Renz, 1946

Fig. 15A–B

Synonymy. – □v [*non*] 1935 *Bulimina spinata* Cushman & Campbell – Cushman & Campbell, p. 72, Pl. 11:11. □v [*non*] 1936b *Bulimina arkadelphiana* Cushman & Parker var. *midwayensis* Cushman & Parker – Cushman & Parker, p. 42, Pl. 7:9–10. □v*1946 *Bulimina petroleana* Cushman & Hedberg var. *spinea* Cushman & Renz – Cushman & Renz, p. 37, Pl. 6:13. □1956 *Bulimina arkadelphiana midwayensis* Cushman & Parker – Said & Kenawy, p. 142, Pl. 4:11. □1983 *Bulimina midwayensis* Cushman & Parker – Tjalsma & Lohmann, p. 6, Pl. 3:1.

□1990 *Bulimina midwayensis* Cushman & Parker – Thomas, Pl. 2:8. □1992 *Praebulimina arkadelphiana* Cushman & Parker – Gawor-Biedowa, p. 111, Pl. 20:11–12; Pl. 21:1. □1992 *Bulimina midwayensis* Cushman & Parker – Kaiho, p. 251, Pl. 3:6. □v1992a *Bulimina spinea* Cushman & Renz – Widmark & Malmgren, p. 110, Pl. 1:5.

Material. – Eighty-six specimens.

Description. – Ovoid with pointing initial end and broadly rounded final end; subcircular to subtriangular in section. Periphery irregular at initial end; smooth at final end. Chambers triserially arranged; chambers of last whorl make up about ½ of the whole test. Sutures indistinct and obscured by ornamentation in earlier part of test; straight, depressed and distinct in the final whorl. Walls calcareous and perforate; ornamentation consists of tiny spines that boarder the base of each chamber, giving the earlier parts of the test an irregular appearance. Aperture narrow and loop-shaped, extending from base of final chamber.

Occurrence. – Maastrichtian and Danian at Sites 525 and 527.

Remarks. – This well-known form is easy to recognize because of its spiny earlier parts of test and its smooth final whorl with distinct and straight sutures. This morphotype is usually reported under the name of *Bulimina midwayensis* Cushman & Parker. The holotypes of three species, all resembling the specimens in the present material, i.e. *Bulimina arkadelphiana* Cushman & Parker var. *midwayensis* Cushman & Parker, *Bulimina spinata* Cushman & Campbell, and *Bulimina petroleana* Cushman & Hedberg var. *spinea* Cushman & Renz, were examined and compared to the specimens found here. The holotype of *Bulimina arkadelphiana* Cushman & Parker var. *midwayensis* Cushman & Parker was found to be separated from the present material in being more sculptured by its costate ornamentation and in its higher degree of inflation of the chambers in the last whorl. Also the holotype of *Bulimina spinata* Cushman & Campbell has a heavier chamber inflation in the last whorl than the specimens in the present material. In addition, the holotype *Bulimina spinata* Cushman & Campbell lacks the apical spine, which is present in most of the specimens analyzed here. Comparison between the holotype of *Bulimina petroleana* Cushman & Hedberg var. *spinea* Cushman & Renz and the specimens found here led, on the other hand, to the conclusion that they may be conspecific, on the basis of the moderate chamber inflation, smooth test walls, distinct, straight sutures in the last whorl, and the spiny initial end, which has an apical spine.

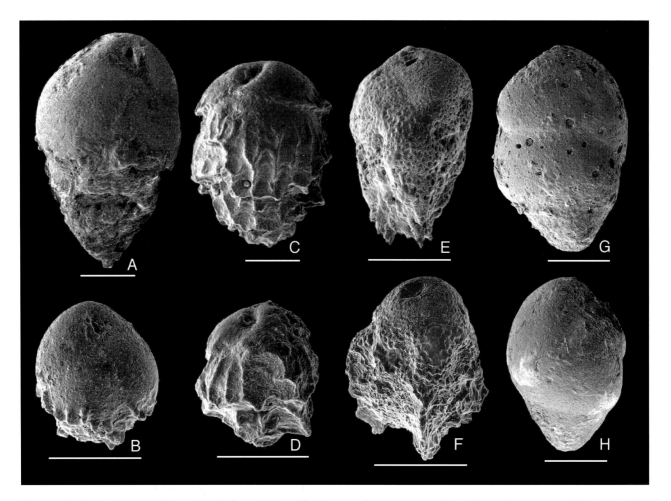

Fig. 15. Scale 100 µm. □A–B. *Bulimina spinea* Cushman & Renz; side views; sample 525A-40-2, 47–48 cm (Maastrichtian). □A. Adult specimen. □B. Juvenile specimen. □C–D. *Bulimina trinitatensis* Cushman & Jarvis; side views; sample 525A-40-1, 149–150 cm (Danian). □C. Adult specimen. □D. Juvenile specimen. □E–F. *Bulimina velascoensis* (Cushman); side views. □E. Adult specimen; sample 527-32-4, 107–108 cm (Maastrichtian). □F. Juvenile specimen; sample 527-32-4, 87–89 cm (Maastrichtian). □G. *Buliminella beaumonti* Cushman & Renz; side view; sample 525A-40-2, 69–70 cm (Maastrichtian). □H. *Buliminella* cf. *beaumonti* Cushman & Renz; side view; sample 525A-40-2, 69–70 cm (Maastrichtian).

Bulimina trinitatensis Cushman & Jarvis, 1928

Fig. 15C–D

Synonymy. – □v*1928 *Bulimina trinitatensis* Cushman & Jarvis, n.sp – Cushman & Jarvis, p. 102, Pl. 14:12. □v1946 *Bulimina trinitatensis* Cushman & Jarvis – Cushman & Renz, p. 37, Pl. 6:8–9. □1973 *Bulimina trinitatensis* Cushman & Jarvis – Douglas, p. 634, Pl. 7:3. □1978 *Bulimina trinitatensis* Cushman & Jarvis – Proto Decima & Bolli, p. 791, Pl. 2:16. □1983 *Bulimina trinitatensis* Cushman & Jarvis – Tjalsma & Lohmann, p. 7, Pl. 3:3–4; Pl. 14:1. □1990 *Bulimina trinitatensis* Cushman & Jarvis – Thomas, p. 589, Pl. 2:7. □1991 *Bulimina trinitatensis* Cushman & Jarvis – Nomura, p. 21, Pl. 1:10. □v1992a *Bulimina trinitatensis* Cushman & Jarvis – Widmark & Malmgren, p. 111, Pl. 1:7. □v1992b *Bulimina trinitatensis* Cushman & Jarvis – Widmark & Malmgren, p. 393, Pl. 1:4.

Material. – 116 specimens.

Description. – Test elongate and subcircular in section. Periphery irregular because of strong ornamentation. Chambers triserially arranged, with a tendency towards bi- or uniseriality in later portion; the lower boarder of chambers extended into a highly ornamented overhanging plate. Sutures distinct because of distinct ornamentation. Walls calcareous; ornamentation organized into numerous longitudinal costae. Aperture loop-shaped, extending from base of last chamber.

Occurrence. – Maastrichtian and Danian at Site 525, Danian at Site 527.

Remarks. – Specimens referred to this very distinct species are consistent with the holotype and some plesiotypes described by Cushman & Jarvis and identified by Cushman & Renz, respectively. The specimens found here also show a close resemblance to the SEM-micrographs of *B. trinitatensis* given by Douglas (1973), Proto Decima & Bolli (1978), and Tjalsma & Lohmann (1983).

Bulimina velascoensis (Cushman, 1925)

Fig. 15E–F

Synonymy. – □v*1925 *Gaudryina velascoensis* n.sp. – Cushman, p. 20, Pl. 3:7. □1929 *Bulimina velascoensis* (Cushman) – White, p. 50, Pl. 5:13. □1977 *Reussella pseudospinulosa* Troelsen – Sliter, p. 675, Pl. 5:8. □[*non*] 1983 *Bulimina velascoensis* (Cushman) – Dailey, p. 766, Pl. 2:13. □1983 *Bulimina velascoensis* (Cushman) – Tjalsma & Lohmann, p. 8, Pl. 3:2. □1986 *Bulimina velascoensis* (Cushman) – Van Morkhoven, p. 335, Pl. 109. □v1992a *Bulimina velascoensis* (Cushman) – Widmark & Malmgren, p. 111, Pl. 1:6. □v1992b *Bulimina velascoensis* (Cushman) – Widmark & Malmgren, p. 393, Pl. 1:4.

Material. – Eight specimens.

Description. – Test oval in outline with a pointing initial end; early portion triangular in section; later portion subtriangular– to–subcircular in section. Periphery of early portion irregular; later portion more even. Chambers triserially arranged, with a tendency towards bi- or uniseriality in later portion; only chambers of last whorl visible. Sutures obscured by ornamentation, except those of the last whorl, which are indistinct and somewhat depressed. Walls calcareous; whole test pitted by pores, which are larger toward the initial end, forming longitudinal costae on the early portion of the test. Aperture loop-shaped, extending from base of last chamber.

Occurrence. – Maastrichtian at Site 527.

Remarks. – Specimens referred to this species correspond very closely to the holotype of *G. velascoensis* Cushman, to the figured specimens of *B. velascoensis* (Cushman) in Tjalsma & Lohmann (1983) and to one of the specimens (pl. 5:8) of *Reussella pseudospinulosa* Troelsen in Sliter (1977).

Family Buliminellidae Hofker, 1951

Diagnosis. – See Loeblich & Tappan (1988, p. 522).

Genus *Buliminella* Cushman, 1911

Type species and diagnosis. – See Loeblich & Tappan (1988, p. 522).

Buliminella beaumonti Cushman & Renz, 1946

Fig. 15G

Synonymy. – □v*1946 *Buliminella beaumonti* n.sp. – Cushman & Renz, p. 36, Pl. 6:7. □1978 *Praebulimina beaumonti* Cushman & Renz – Proto Decima & Bolli, p. 795, Pl. 1:19. □1983 *Buliminella beaumonti* Cushman & Renz – Dailey, p. 766, Pl. 2:3. □1983 *Buliminella beaumonti* Cushman & Renz Tjalsma & Lohmann, p. 9, Pl. 3:6. □v1988 *Buliminella? beaumonti* Cushman & Renz – Widmark & Malmgren, p. 67, Pl. 1:3. □v1992a *Buliminella? beaumonti* Cushman & Renz – Widmark & Malmgren, p. 111, Pl. 1:9.

Material. – Thirty-six specimens.

Description. – Test ovate and subcircular in section. Periphery rounded. Chambers arranged in a trochospiral coil, about four inflated chambers in the final whorl. Axis of coiling straight. Suture distinct and translucent; sutures sinuous around apertural face; lobated at base of chambers (at least in final whorl). Walls calcareous, smooth, and finely perforated (giving the test a lustrous appearance). Aperture subcircular, located centrally in apertural view, at the base of final chamber; aperture has a toothplate.

Occurrence. – Maastrichtian and Danian at Site 525, Maastrichtian at Site 527.

Remarks. – The holotype and paratypes of this species have a higher number of sutural lobes (holotype: 5–8 lobes per chamber) than the specimens referred to this species herein. Furthermore, distinction between specimens belonging to *Buliminella* on one hand and *Quadratobuliminella* on the other was difficult. However, specimens with less inflated chambers (= more regular test shape and subcircular transection) were referred to *Buliminella*, whereas specimens with more inflated chambers (= more irregular test shape and subquadrate transection) were referred to *Quadratobuliminella*. Only forms with distinct, sinuous, and translucent sutures were included in this species (compare *B.* cf. *beaumonti* below).

Buliminella cf. *beaumonti* Cushman & Renz, 1946

Fig. 15H

Synonymy. – □See *B. beaumonti* above. □v1994 *Sitella cushmani* (Sandidge) – Speijer, p. 50, Pl. 1:9.

Material. – 163 specimens.

Description. – Test ovate and subcircular in section. Periphery rounded. Chambers trochospirally arranged, about four inflated chambers in the final whorl. Axis of coiling straight. Sutures distinct and translucent; sutures sinuous around apertural face; curved at base of chambers. Walls calcareous, smooth, and finely perforated (giving the test a lustrous appearance). Aperture subcircular, located centrally in apertural view, at the base of final chamber; aperture has a toothplate.

Occurrence. – Maastrichtian and Danian at Site 525 and Site 527.

Remarks. – This taxon is very similar to *B. beaumonti* distinguished here, but lacks the typical sutures of *B. beaumonti*, which are sinuous and lobated. Specimens close to *B. grata* Parker & Bermudez are also included in this taxon.

Buliminella cf. *plana* Cushman & Parker, 1936

Fig. 16A

Synonymy. – □v*1936a cf. *Buliminella carseyae* Plummer var. *plana* Cushman & Parker – Cushman & Parker, p. 8, Pl. 2:7. □v1970 cf. *Buliminella carseyae plana* Cushman & Parker – Todd, p. 140, Pl. 1:16. □v1992b *Buliminella* cf. *plana* Cushman & Parker – Widmark & Malmgren, p. 393, Pl. 1:2.

Material. – Thirty-six specimens.

Description. – Test ovate and subcircular in section. Periphery rounded. Chambers trochospirally arranged and slightly inflated, about four chambers in the final whorl; the last one with a large, flattened apertural face. Coiling axis slightly bent. Sutures distinct and translucent; sutures weakly sinuous around apertural face; curved at base of chambers. Walls calcareous, smooth, and finely perforated (giving the test a lustrous appearance). Aperture subcircular, located centrally in apertural view, at the base of final chamber; aperture has a toothplate.

Occurrence. – Maastrichtian at Sites 525 and 527.

Remarks. – Specimens of this taxon are very similar to the specimens identified as *Buliminella carseyae plana* Cushman & Parker by Todd (1970). Both the specimens of *B.* cf. *plana* in the present material and Todd's (1970) specimens of *B. carseyae plana* have a large, flattened apertural face in common, which could be used for consistent separation of this form from other buliminids in the present material. However, the holotype of *Buliminella carseyae* Plummer var. *plana* Cushman & Parker differs from the specimens of *B.* cf. *plana* in the present material and Todd's (1970) specimens of *B. carseyae plana* in having a more fusiform test shape and narrower sutures (like many

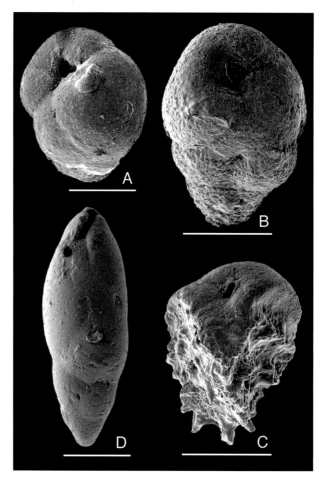

Fig 16. Scale 100 µm. □A. *Buliminella* cf. *plana* Cushman & Parker; side view; sample 527-32-4, 57–59 cm (Maastrichtian). □B. *Quadratobuliminella* spp.; side view; sample 525A-40-2, 78–79 cm (Maastrichtian). □C. *Reussella szajnochae* (Grzybowski); side view; sample 527-32-6, 30–31 cm (Maastrichtian). □D. *Fursenkoina?* spp.; side view; sample 525A-40-2, 109–110 cm (Maastrichtian).

Praebulimina), and in having a less flattened apertural face. This form is separated from similar forms on the basis of its flat apertural face and slightly bent coiling axis.

Genus *Quadratobuliminella* de Klasz, 1953

Type species and diagnosis. – See Loeblich & Tappan (1988, p. 522).

Quadratobuliminella spp.

Fig. 16B

Material. – Thirty-three specimens.

Occurrence. – Maastrichtian and Danian at Site 525, Maastrichtian at Site 527.

Remarks. – This taxon is very similar to *Buliminella beaumonti* Cushman & Renz. The topotype of *Q. pyramidalis* de Klasz (USNM No. 433512), figured by Loeblich & Tappan (1964), was examined and found to be more regularly quadrate in transection than the specimens encountered in the present material. The specimens found here are thus morphologically intermediate between *B. beaumonti* and *Q. pyramidalis*. Hence, specimens with numerous, inflated chambers and irregular test shapes (somewhat subquadrate in section) are included in this taxon.

Family Reussellidae Cushman, 1933

Diagnosis. – See Loeblich & Tappan (1988, p. 526).

Genus *Reussella* Galloway, 1933

Type species and diagnosis. – See Loeblich & Tappan (1988, p. 527).

Reussella szajnochae (Grzybowski, 1896)

Fig. 16C

Synonymy. – □*1896 *Verneuilina szajnochae* n.sp. – Grzybowski, p. 287, Pl. 9:19. □1977 *Reussella szajnochae* (Grzybowski) – Sliter, p. 682, Pl. 5:9–10. □1978 *Reussella szajnochae* (Grzybowski) – Beckmann, p. 768, Pl. 2:16–17. □v1988 *Reussella szajnochae* (Grzybowski) – Widmark & Malmgren, p. 69, Pl. 1:6. □1992 *Pyramidina szajnochae* (Grzybowski) – Gawor-Biedowa, p. 123, Pl. 24:8–10. □v1992a *Reussella szajnochae* (Grzybowski) – Widmark & Malmgren, p. 113, Pl. 1:15. □v1992b *Reussella szajnochae* (Grzybowski) – Widmark & Malmgren, p. 402, Pl. 1:7.

Material. – Twenty-four specimens.

Description. – Test tapering and triangular in section; sides of test distinctly concave. Periphery sharply angular and spinose. Chambers triserially arranged; margins of chambers have large irregular spines. Sutures distinct, broad, slightly curved, and elevated. Walls calcareous, coarsely perforate, and ornamented by spines. Aperture basal in final chamber, consisting of a low arch.

Occurrence. – Maastrichtian at Sites 525 and 527 (a single specimen was encountered in one of the Danian samples from Site 527).

Remarks. – This well-known species is easy to recognize by its spiny outgrowths at the edges of its triangular test.

Superfamily Fursenkoinacea Loeblich & Tappan, 1961

Diagnosis. – See Loeblich & Tappan (1988, p. 529).

Family Fursenkoinidae Loeblich & Tappan, 1961

Diagnosis. – See Loeblich & Tappan (1988, p. 529).

Genus *Fursenkoina* Loeblich & Tappan, 1961

Type species and diagnosis. – See Loeblich & Tappan (1988, p. 530).

Fursenkoina? spp.

Fig. 16D

Material. – Five specimens.

Occurrence. – Maastrichtian at Site 525, Danian at Site 527.

Remarks. – A few specimens with twisted biserial chamber arrangements, narrow tests, and terminal, comma-shaped apertures, were referred to this taxon.

Superfamily Pleurostommellacea Reuss, 1860

Diagnosis. – See Loeblich & Tappan (1988, p. 535).

Family Pleurostommellidae Reuss, 1860

Diagnosis. – See Loeblich & Tappan (1988, p. 535).

Subfamily Pleurostommellinae Reuss, 1860

Diagnosis. – See Loeblich & Tappan (1988, p. 535).

Genus *Ellipsobulimina* Silvestri, 1903

Type species and diagnosis. – See Loeblich & Tappan (1988, p. 536).

Ellipsobulimina? sp.

Fig. 17A

Material. – One specimen.

Occurrence. – Danian at Site 257.

Remarks. – A single rotaliid specimen with ovate test shape and a terminal, semilunate aperture was assigned to this taxon.

Genus *Ellipsodimorphina* Silvestri, 1901

Type species and diagnosis. – See Loeblich & Tappan (1988, p. 536).

Ellipsodimorphina? spp.

Fig. 17B

Material. – Five specimens.

Occurrence. – Maastrichtian at Site 525, Danian and Maastrichtian at Site 527.

Remarks. – A few rotaliid specimens with elongate tests were assigned to this taxon. They are all characterized by an early biserial stage, which abruptly changes into a later uniserial portion. The aperture is formed as an elongate, arched slit.

Genus *Ellipsoglandulina* Silvestri, 1900

Type species and diagnosis. – See Loeblich & Tappan (1988, p. 536).

Ellipsoglandulina? spp.

Fig. 17C

Material. – Seven specimens.

Occurrence. – Maastrichtian at Site 525, Danian and Maastrichtian at Site 527.

Remarks. – A few rotaliid specimens, characterized by strongly overlapping chambers, which are uniserially arranged, and a terminal semilunate aperture, were assigned to this taxon.

Genus *Ellipsoidella* Heron-Allen & Earland, 1910

Type species and diagnosis. – See Loeblich & Tappan (1988, p. 537).

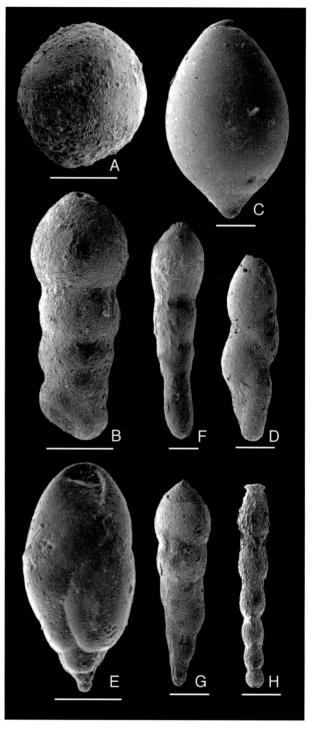

Fig. 17. Scale 100 µm. □A. *Ellipsobulimina*? spp.; side view; sample 527-32-4, 39–40 cm (Danian). □B. *Ellipsodimorphina*? spp.; side view; sample 525A-40-2, 59–60 cm (Maastrichtian). □C. *Ellipsoglandulina*? spp.; side view; sample 527-32-5, 57–58 cm (Maastrichtian). □D. *Ellipsoidella*? spp.; side view; sample 525A-40-2, 92–93 cm (Maastrichtian). □E. *Ellipsopolymorphina*? spp.; side view; sample 527-32-4, 19–20 cm (Danian). □F. *Nodosarella* spp.; side view; sample 527-32-3, 129–130 cm (Danian). □G. *Pleurostomella* spp.; side view; sample 527-32-3, 39–40 cm (Danian). □H. *Nodogenerina* spp.; side view; sample 527-32-3, 129–130 cm (Danian).

Ellipsoidella? spp.

Fig. 17D

Material. – Eleven specimens.

Occurrence. – Maastrichtian at Site 525, Danian and Maastrichtian at Site 527.

Remarks. – A few rotaliid specimens with elongate tests were assigned to this taxon. They are characterized by an early biserial stage, which slowly becomes uniserial in the later portion of test. The subterminal aperture is formed as an elongate, arched slit.

Genus *Ellipsopolymorphina* Silvestri, 1901

Type species and diagnosis. – See Loeblich & Tappan (1988, p. 537).

Ellipsopolymorphina? spp.

Fig. 17E

Material. – Eight specimens.

Occurrence. – Maastrichtian at Site 525, Danian and Maastrichtian at Site 527.

Remarks. – A few rotaliid specimens with elongate tests were assigned to this taxon. They are characterized by strongly overlapping chambers, which are biserially arranged in early stage and become uniserial in the later portion of the test. The terminal aperture is formed as a semilunate slit.

Genus *Nodosarella* Rzehak, 1895

Type species and diagnosis. – See Loeblich & Tappan (1988, p. 537).

Nodosarella spp.

Fig. 17F

Material. – Seven specimens.

Occurrence. – Maastrichtian at Site 525, Danian at Site 527.

Remarks. – A few rotaliid specimens with elongate tests were assigned to this taxon. They are characterized by their inflated chambers, which are uniserially arranged. The terminal aperture is formed as a slit.

Genus *Pleurostomella* Reuss, 1860

Type species and diagnosis. – See Loeblich & Tappan (1988, p. 538).

Pleurostomella spp.

Fig. 17G

Material. – Eighteen specimens.

Occurrence. – Maastrichtian at Site 525, Danian and Maastrichtian at Site 527.

Remarks. – A few rotaliid specimens with elongate tests were assigned to this taxon. They are characterized by an early biserial stage, which becomes uniserial in the later portion of test. *Pleurostomella* spp. differs from *Ellipsoidella*? spp. in the nature of the sutures; the sutures in *P.* spp. become more straight (nearly horizontal) in the later uniserial part of test.

Superfamily Stilostomellacea Finlay, 1947

Diagnosis. – See Loeblich & Tappan (1988, p. 539).

Family Stilostomellidae Finlay, 1947

Diagnosis. – See Loeblich & Tappan (1988, p. 539).

Genus *Nodogenerina* Cushman, 1927

Type species and diagnosis. – See Loeblich & Tappan (1988, p. 539).

Nodogenerina spp.

Fig. 17H

Material. – Eight specimens.

Occurrence. – Maastrichtian at Site 525, Danian at Site 527.

Remarks. – A few specimens, mostly fragmented, with uniserial, rectilinear tests, subglobular chambers, and terminal apertures, were referred to this taxon.

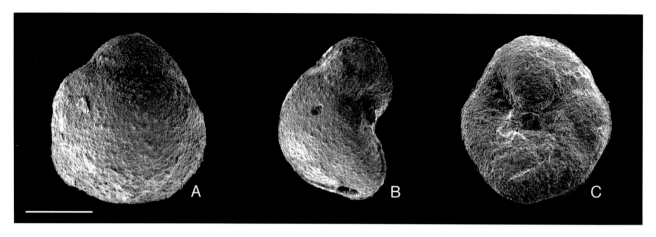

Fig. 18. Scale 100 μm. □A–C. *Rosalina*? sp.; sample 527-32-4, 0–1 cm (Danian). □A. Spiral view. □B. Peripheral view. □C. Umbilical view.

Superfamily Discorbacea Ehrenberg, 1838

Diagnosis. – See Loeblich & Tappan (1988, p. 541).

Family Rosalinidae Reiss, 1963

Diagnosis. – See Loeblich & Tappan (1988, p. 560).

Genus *Rosalina* d'Orbigny, 1826

Type species and diagnosis. – See Loeblich & Tappan (1988, p. 561).

Rosalina? spp.

Fig. 18A–C

Material. – Thirty-two specimens.

Occurrence. – Maastrichtian at Site 525, Danian and Maastrichtian at Site 527.

Remarks. – The taxonomic status of this taxon is uncertain. A small number of minute specimens with a pronounced convex spiral side and a concave umbilical side was assigned to this taxon.

Superfamily Discorbinellacea Sigal, 1952

Diagnosis. – See Loeblich & Tappan (1988, p. 572).

Family Parrelloididae Hofker, 1956

Diagnosis. – See Loeblich & Tappan (1988, p. 572).

Genus *Cibicidoides* Thalmann, 1939

Type species and diagnosis. – See Loeblich & Tappan (1988, p. 572).

Cibicidoides dayi (White, 1928)

Fig. 19A–F

Synonymy. – □*1928b *Planulina dayi* n.sp. – White, p. 300, Pl. 41:3. □1983 *Cibicidoides dayi* (White) – Dailey, p. 766, Pl. 9:1–3. □1983 *Gavelinella monterelensis* (Marie) – Dailey, p. 766, Pl. 8:13. □1983 *Gavelinella stephensoni* (White) – Dailey, p. 767, Pl. 10:5–6,9. □1983 *Cibicidoides dayi* (White) – Tjalsma & Lohmann, p. 9, Pl. 6:7. □1986 *Cibicidoides dayi* (White) – Van Morkhoven *et al.*, p. 353, Pl. 114. □1991 *Cibicidoides dayi* (White) – Nomura, p. 22, Pl. 4:1. □?1991 *Gavelinella* sp. – Nomura, p. 22, Pl. 4:5.

Material. – Twenty-two specimens.

Description. – Test trochospiral, biconvex, and biumbonate; lenticular in transection. Periphery subacute without hyaline border. Aperture consisting of an equatorial slit, extending from periphery and along spiral suture of the half last whorl, forming a distinct groove. Walls calcareous and opaque; coarsely perforated, especially in depressed areas between sutures dorsally, less perforated ventrally. Sutures indistinct, broad, curved, elevated dorsally and flush ventrally. Dorsal side (spiral side) convex and not completely evolute, with umbonal plug. All dorsal chambers indistinctly visible; chambers increase very slowly in size as added, they are very narrow and numer-

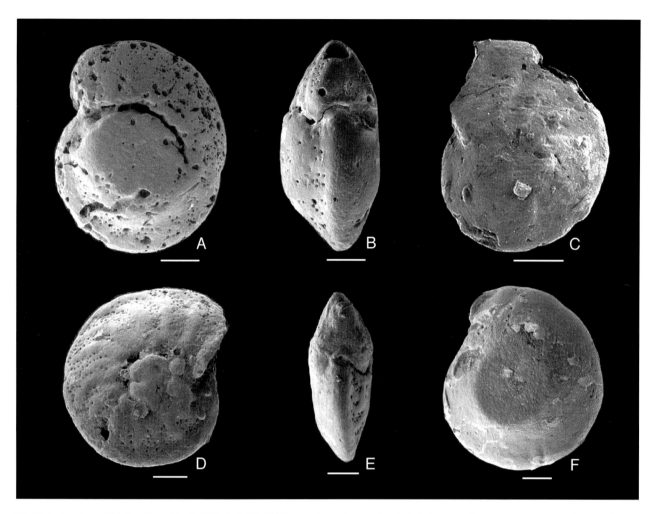

Fig. 19. Scale 100 μm. □A–F. *Cibicidoides dayi* (White). □A–C. 'Biconvex' morphotype. □A. Spiral view; sample 525A-40-4, 60–61 cm (Maastrichtian). □B. Peripheral view; sample 525A-40-2, 128–129 cm (Maastrichtian). □C. Umbilical view; sample 525A-40-4, 60–61 cm (Maastrichtian). □D–F. 'Lenticular' morphotype; sample 525A-40-2, 69–70 cm (Maastrichtian). □D. Spiral view. □E. Peripheral view. □F. Umbilical view.

ous (about 15–20 in the final whorl). Ventral side (umbilical side) convex and involute, with a flattened umbo instead of umbilicus. Ventral chambers indistinct, and only those of final whorl visible.

Occurrence. – Danian and Maastrichtian at Site 525, Maastrichtian at Site 527.

Remarks. – The original concept of this species is not consistent; both Tjalsma & Lohmann (1983) and Van Morkhoven *et al.* (1986) concluded after examination of the primary types of this species that some of the paratypes belong to Fisher's (1969) species *hyphalus* (= *Cibicidoides hyphalus*). It was necessary to apply a relatively wide concept for this species in order to achieve a consistent quantification of the material. *Cibicidoides dayi* is here considered to be represented by two morphotypes with overlapping morphologies: a more strongly 'biconvex' morphotype (Fig. 19A–C) on the one hand, and a less

biconvex or 'lenticular' morphotype (Fig. 19D–F), on the other. The 'biconvex' morphotype has a prominent evolute, convex dorsal side and a large dorsal umbonal plug, in contrast to the 'lenticular' morphotype, which is more compressed dorsally–ventrally (lenticular in transection), has a less evolute dorsal side, and a smaller dorsal umbonal plug. *Cibicidoides dayi* is here separated from other *Cibicidoides* in the material by its larger size, opaque test walls, and more or less equally biconvex test shape. The figured specimens of *Gavelinella stephensoni* (White) given by Dailey (1983) are very close to the 'biconvex' morphotype, as well as *Cibicidoides dayi* (White) figured in the same source. The 'lenticular' morphotype resembles the specimens of *Gavelinella monterelensis* (Marie) figured by Dailey (1983; Pl. 8:13) and of *Cibicidoides dayi* (White) figured by Tjalsma & Lohmann (1983; Pl. 6:7) with respect to the weakly evolute coiling and the small umbilical plug on the dorsal side.

Cibicidoides hyphalus (Fisher, 1969)

Fig. 20A–E

Synonymy. – □*1969 *Anomalinoides hyphalus* sp. nov. – Fisher, p. 197, Text-fig. 3. □1983 *Gavelinella hyphalus* (Fisher) – Dailey, p. 766, Pl. 10:1–3. □1983 *Gavelinella hyphalus* (Fisher) – Tjalsma & Lohmann, p. 13, Pl. 4:8–9; Pl. 7:11. □1986 *Cibicidoides hyphalus* (Fisher) – Van Morkhoven *et al.*, p. 359, Pl. 116. □*v*1988 *Gavelinella hyphalus* (Fisher) – Widmark & Malmgren, p. 76, Pl. 5:2. □1991 *Cibicidoides hyphalus* (Fisher) – Nomura, p. 22, Pl. 3:9. □*v*1992a *Cibicidoides hyphalus* (Fisher) – Widmark & Malmgren, p. 111, Pl. 5:1–2. □*v*1992b *Cibicidoides hyphalus* (Fisher) – Widmark & Malmgren, p. 393, Pl. 2:10–11.

Material. – About 640 specimens.

Description. – Test trochospiral, planoconvex to unequally biconvex, and biumbonate. Periphery rounded to weakly angular, with a narrow hyaline boarder. Aperture consisting of an equatorial slit, extending from periphery and along spiral suture, forming a groove. Walls calcareous and transparent; dorsally perforated by pores located in depressed area between the sutures; ventrally perforated and smooth. Dorsal side (spiral side) flat to weakly convex and evolute, with umbonal plug. All dorsal chambers visible, narrow and slowly increasing in size as added; about ten chambers in final whorl; previous chambers obscured by hyaline plug or by triangular flap-like extensions of later formed chambers; extensions sometimes fused into an irregular umbilical plug. Dorsal sutures indistinct, broad, curved, and elevated. Ventral side (umbilical side) convex (more than dorsally) and involute; umbilicus closed by hyaline cover. Only chambers of final whorl (10–12 in number) visible ventrally; chambers of previous whorls sometimes visible through hyaline cover. Ventral sutures distinct, broad, curved, and flush.

Occurrence. – Danian and Maastrichtian at Sites 525 and 527.

Remarks. – This species is distinguished by its nearly flat, almost evolute spiral side, which has numerous large pores, and its convex involute umbilical side, which has numerous chambers. Two morphotypes, with overlapping morphologies, represents this species in the present material. One is somewhat lenticular in side view, has a subacute periphery, and a well-defined dorsal umbonal plug outlined by the spiral sutural aperture, a feature which is characteristic for the genus of *Cibicidoides* (Fig. 20A–B). The other morphotype has a rounded periphery

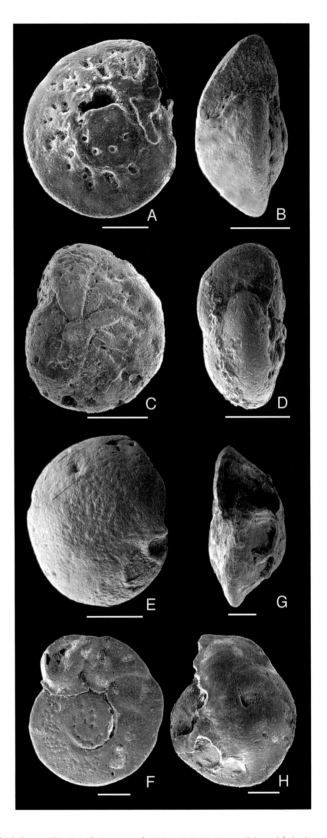

Fig. 20. Scale 100 μm. □A–E. *Cibicidoides hyphalus* (Fisher). □A–B. '*Cibicidoides*'-type. □A. Spiral view; sample 525A-40-2, 59–60 cm (Maastrichtian). □B. Peripheral view; sample 525A-40-2, 69–70 cm (Maastrichtian). □C–D. '*Gavelinella*'-type; sample 525A-40-2, 100–102 cm (Maastrichtian). □C. Spiral view. □D. Peripheral view. □E. Umbilical view; sample 525A-40-4, 60–61 cm (Maastrichtian). □F–H. *Cibicidoides* sp. planoconvex form. □F. Spiral view; sample 525A-40-2, 119–120 cm (Maastrichtian). □G. Peripheral view; sample 525A-40-2, 92–93 cm (Maastrichtian). □H. Umbilical view; sample 525A-40-2, 92–93 cm (Maastrichtian).

and a weakly developed umbonal plug (Fig. 20C–D) and has often been referred to *Gavelinella* in the literature. *Cibicidoides hyphalus* is previously reported from the Maastrichtian of the South Atlantic by Dailey (1983), who referred it to *Gavelinella*.

Cibicidoides sp. planoconvex form

Fig. 20F–H

Synonymy. – □1983 *Gavelinella eriksdalensis* (Brotzen) – Dailey, p. 766, Pl. 9:5,9–11. □1991 *Cibicidoides* sp. 5 – Nomura, p. 22, Pl. 4:4. □v1994 *Cibicidoides suzakensis* (Bykova) – Speijer, p. 54, Pl. 5:1.

Material. – Fifty-one specimens.

Description. – Test trochospiral, biumbonate, and plano-convex. Periphery acute with narrow, hyaline keel. Aperture consisting of an equatorial slit, extending from periphery and along spiral suture of the last two or three chambers, forming a distinct groove. Walls calcareous; dorsally coarsely perforated, except for central (umbonal) area, which consists of a thickening of hyaline material through which previous whorls are visible; ventrally smooth and more finely perforated than dorsally. Dorsal side (spiral side) flat to weakly convex and evolute; adult specimens having a flatter dorsal side than juveniles, which are more biconvex. All dorsal chambers visible; about ten chambers in final whorl, gradually increasing in size as added; chambers of inner whorls obscured by hyaline umbo. Dorsal sutures curved, distinct, and depressed between the two or three last formed chambers in the adult specimens; indistinct and flush between previous chambers and in the juvenile specimens. Ventral side (umbilical side) strongly convex and involute; umbilicus substituted by a hyaline boss. Only chambers of final whorl visible ventrally. Ventral sutures distinct, broad, and curved to weakly sinuous; sutures depressed in adult specimens; flush in the juvenile.

Occurrence. – Danian and Maastrichtian at Site 525.

Remarks. – This relatively stable form is separated from other *Cibicidoides* encountered in the material by its planoconvex test and distinct hyaline boss on the ventral side. They are very close to the figured specimens of *Gavelinella eriksdalensis* (Brotzen) in Dailey (1983). However, syntypes of *Cibicides eriksdalensis* Brotzen, examined at the Smithsonian Institution, do not have a ventral hyaline boss, which is characteristic for the taxon found here (the umbilical area is instead marked by a large, deep umbilicus in the syntypes). In addition, the spiral side is much more irregular and the umbilical side is less convex in the syntypes than in the specimens found in the present material.

Cibicidoides velascoensis (Cushman, 1925)

Fig. 21A–C

Synonymy. – □v*1925 *Anomalina velascoensis* n.sp. – Cushman p. 21, Pl. 3:3. □1928b *Planulina velascoensis* (Cushman) – White, p. 303, Pl. 41:7. □1977 *Gavelinella velascoensis* (Cushman) – Sliter, p. 675, Pl. 13:1. □1983 *Gavelinella velascoensis* (Cushman) – Dailey, p. 767, Pl. 9:12–13. □1983 *Gavelinella velascoensis* (Cushman) – Tjalsma & Lohmann, p. 14, Pl. 5:8. □1986 *Cibicidoides velascoensis* (Cushman) – Van Morkhoven *et al.*, p. 371, Pl. 121. □1991 *Cibicidoides velascoensis* (Cushman) – Nomura, p. 22, Pl. 3:5. □v1992a *Cibicidoides velascoensis* (Cushman) – Widmark & Malmgren, p. 111, Pl. 5:5.

Material. – 114 specimens.

Description. – Test trochospiral, nearly planoconvex; sub-circular in outline. Periphery broadly rounded. Aperture consisting of an equatorial slit, extending from periphery and along spiral suture, forming an irregular groove. Walls calcareous, coarsely perforated and rough by ridges between sutures dorsally, and finely perforate and smooth ventrally. Dorsal side (spiral side) flat or weakly convex and involute; umbo partly filled by hyaline material. Only chambers of last whorl visible dorsally. Dorsal sutures distinct, broad, curved, and depressed. Ventral side (umbilical side) broadly convex and evolute, with a tendency towards involute coiling. Ventral chambers of final whorl visible, whereas chambers of previous whorls are obscured; chambers relatively narrow and about nine in number. Ventral sutures flush and curved, indistinct in adult specimens.

Occurrence. – Danian and Maastrichtian at Sites 525 and 527.

Remarks. – This large species is easy to recognize on the basis of its broadly rounded periphery, ridges and large pores dorsally, and ventrally incomplete involute coiling. It differs from *C. hyphalus*, which has a less convex, involute ventral side. The present material is conspecific with the holotype of this species, which was available and examined at the Cushman Collection, Smithsonian Institution.

Superfamily Planorbulinacea Schwager, 1877

Diagnosis. – See Loeblich & Tappan (1988, p. 579).

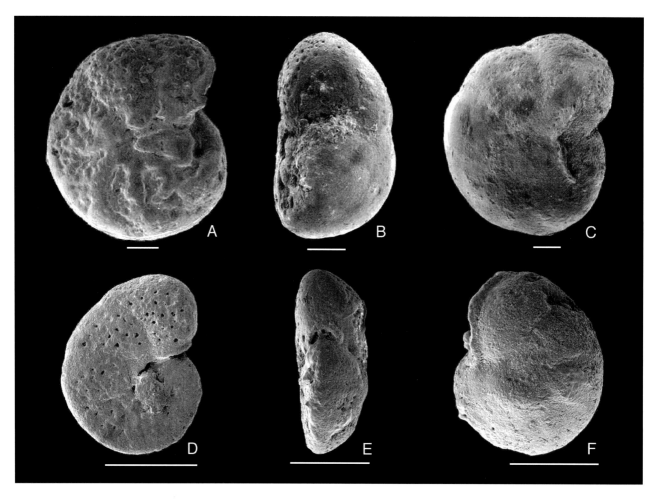

Fig. 21. Scale 100 μm. □A–C. *Cibicidoides velascoensis* (Cushman). □A. Spiral view; sample 525A-40-2, 47–48 cm (Maastrichtian). □B. Peripheral view; sample 525A-40-2, 47–48 cm (Maastrichtian). □C. Umbilical view; sample 525A-40-2, 92–93 cm (Maastrichtian). □D–F. *Cibicides* cf. *excavatus* Brotzen; sample 525A-40-2, 119–120 cm (Maastrichtian). □D. Spiral view. □E. Peripheral view. □F. Umbilical view.

Family Cibicididae Cushman, 1927

Diagnosis. – See Loeblich & Tappan (1988, p. 581).

Subfamily Cibicidinae Cushman, 1927

Diagnosis. – See Loeblich & Tappan (1988, p. 581).

Genus *Cibicides* de Montfort, 1808

Type species and diagnosis. – See Loeblich & Tappan (1988, p. 582).

Cibicides cf. *excavatus* Brotzen, 1936

Fig. 21D–F

Synonymy. – □cf. *1936 *Cibicides excavata* n.sp. – Brotzen, p. 189, Pl. 13:7–8. □1983 *Cibicides excavatus* Brotzen – Dailey, p. 766, Pl. 3:5–7.

Material. – Eight specimens.

Description. – Test small, trochospiral, and planoconvex. Periphery subangular. Aperture consisting of an interiomarginal slit that extends from periphery along spiral suture for a short distance. Walls calcareous and smooth; coarsely perforated dorsally and more finely perforated ventrally. Dorsal side (spiral side) flat to weakly concave and evolute. Chambers hardly visible dorsally and arranged in one and a half whorls only; about seven chambers in the final whorl; chambers increasing gradually in size as added. Dorsal sutures indistinct, curved, and flush. Ventral side (umbilical side) convex and involute;

umbilicus closed. Only chambers of final whorl visible ventrally. Ventral sutures distinct, broad, curved, and slightly depressed.

Occurrence. – Danian and Maastrichtian at Site 525.

Remarks. – This taxon occurs only in one Danian sample and a few of the Maastrichtian samples. The planoconvex test, apertural features, and chamber arrangement are similar to those of the specimens of *Cibicides excavatus* Brotzen figured by Dailey (1983). Only two specimens referred to *Cibicides excavata* Brotzen were available at the Naturhistoriska Riksmuseet. They were collected from the type locality of this species (Eriksdal, Sweden) and identified by Brotzen and thus regarded as good representatives of *Cibicides excavata* Brotzen. They differ, however, from the specimens in the present material in being more convex ventrally and more excavated dorsally, and in having a heavier chamber inflation in the last whorl.

Superfamily Asterigerinacea d'Orbigny, 1839

Diagnosis. – See Loeblich & Tappan (1988, p. 600).

Family Epistomariidae Hofker, 1954

Diagnosis. – See Loeblich & Tappan (1988, p. 600).

Subfamily Nuttallidinae Saidova, 1981

Diagnosis. – See Loeblich & Tappan (1988, p. 602).

Genus *Nuttallides* Finlay, 1939

Type species and diagnosis. – See Loeblich & Tappan (1988, p. 603).

Nuttallides sp. A
Fig. 22A–C

Synonymy. – □*v1988 Nuttallides* sp. a – Widmark & Malmgren, p. 69, Pl. 2:1. □*v1992a Nuttallides* sp. A – Widmark & Malmgren, p. 112, Pl. 2:1. □*v1992b Nuttallides* sp. A – Widmark & Malmgren, p. 394, Pl. 5:1–3.

Material. – 166 specimens.

Description. – Test trochospiral, biconvex, and lenticular in section; subcircular and lobated in outline. Periphery sharp with a narrow hyaline boarder. Walls calcareous, smooth, and finely perforate. Aperture consisting of an interiomarginal, elongate slit, extending from periphery to umbilical boss. Dorsal side (spiral side) slightly convex and evolute. All chambers visible on dorsal side, about six in number in final whorl; chambers longer than wide (ratio 3:1) and arranged into three whorls. Dorsal sutures distinct, narrow, flush, oblique, and slightly curved backwards. Ventral side (umbilical side) convex and involute; instead of umbilicus a weakly developed hyaline boss of glassy material. Only chambers of last whorl visible on ventral side; 6–7 in number. Ventral sutures distinct, radiate, narrow, and curved.

Occurrence. – Maastrichtian and Danian at Site 527.

Remarks. – This form resembles *Nuttallides truempyi* (Nuttall) but is more equally biconvex in peripheral view and has a less developed umbilical boss and curved sutures rather than sinuous ones on the umbilical side.

Nuttallides sp. B
Fig. 22D–F

Synonymy. – □*v1992a Nuttallides* sp. B – Widmark & Malmgren, p. 112, Pl. 2:2.

Material. – 122 specimens.

Description. – Test trochospiral, biconvex, lenticular in section; outline circular and less lobated than in other *Nuttallides* described here. Periphery sharp with narrow hyaline boarder. Walls calcareous, smooth, and finely perforate. Aperture consisting of an interiomarginal, elongate slit, extending from periphery to umbilical boss. Dorsal side (spiral side) slightly convex and evolute. All chambers visible on dorsal side, numerous, about ten in number in final whorl; chambers longer than wide (ratio 2:1); 3–4 whorls; chambers slowly increasing in size as added. Dorsal sutures distinct, narrow, flush, nearly perpendicular to spiral net, and somewhat curved. Ventral side (umbilical side) strongly convex and involute; umbilicus substituted by a well-developed hyaline boss of glassy material, through which earlier whorls are visible. Only chambers of last whorl visible on ventral side; 9–10 in number. Ventral sutures distinct, radiate, more or less straight, depressed, and bordering toward the umbilical boss.

Occurrence. – Maastrichtian and Danian at Site 527.

Remarks. – Specimens with a well-developed umbonal boss and a strongly convex umbilical side were assigned to this species. *Nuttallides* sp. B is different from *N.* sp. A and *N. truempyi* (Nuttall) in having straight umbilical sutures (instead of curved or sinuous). In addition, *Nuttallides* sp. B has chambers on the umbilical side that are narrower and more numerous than in *N. truempyi* and *N.* sp. A.

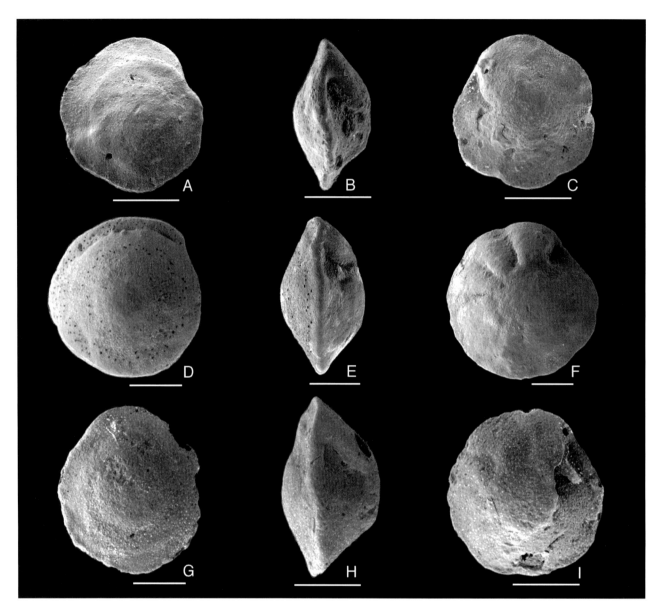

Fig. 22. Scale 100 µm. □A–C. *Nuttallides* sp. A. □A. Spiral view; sample 527-32-4, 87–89 cm (Maastrichtian). □B. Peripheral view; sample 527-32-4, 107–108 cm (Maastrichtian). □C. Umbilical view; sample 527-32-4, 107–108 cm (Maastrichtian). □D–F. *Nuttallides* sp. B. □D. Spiral view; sample 527-32-4, 87–89 cm (Maastrichtian). □E. Peripheral view; sample 527-32-4, 87–89 cm (Maastrichtian). □F. Umbilical view; sample 527-32-4, 107–108 cm (Maastrichtian). □G–I. *Nuttallides truempyi* (Nuttall); sample 525A-40-2, 59–60 cm (Maastrichtian). □G. Spiral view. □H. Peripheral view. □I. Umbilical view.

Nuttallides truempyi (Nuttall, 1930)

Fig. 22G–I

Synonymy. – □*1930 *Eponides truempyi* n.sp. – Nuttall, p. 287, Pl. 3:5. □v1935 *Astigerina crassaformis* n.sp. – Cushman & Siegfus, p. 94, Pl. 10:14. □v1946 *Eponides bronnimanni* n.sp. – Cushman & Renz, p. 45, Pl. 7:24. □1973 *Nuttallides truempyi* (Nuttall) – Douglas, Pl. 21:4–5, p. 668, Pl. 24:1–2. □1977 *Osangularia cordieriana* (d'Orbigny) – Sliter, p. 675, Pl. 9:1–3,9. □1978 *Nuttallides bronnimanni* (Cushman & Renz) – Beckmann, p. 768, Pl. 3:1–2. □1983 *Nuttallides truempyi* (Nuttall) – Dailey, p. 767, Pl. 3:2–3. □1983 *Nuttallides truempyi* (Nuttall) – Tjalsma & Lohmann, p. 17, Pl. 6:4; Pl. 17:4–5; Pl. 21:1–4. □1986 *Nuttallides truempyi* (Nuttall) – Van Morkhoven *et al.*, p. 288, Pl. 96. □v1988 *Nuttallides truempyi* (Nuttall) – Widmark & Malmgren, p. 69, Pl. 1:7. □1991 *Nuttallides truempyi* (Nuttall) – Nomura, p. 22, Pl. 2:7. □v1992a *Nuttallides truempyi* (Nuttall) – Widmark & Malmgren, p. 112, Pl. 2:3. □v1992b *Nuttallides truempyi* (Nuttall) – Widmark & Malmgren, p. 398, Pl. 5:4–6.

Material. – About 1100 specimens.

Description. – Test trochospiral, equally to unequally biconvex, and lenticular in section; circular and lobated in outline. Periphery sharp with narrow hyaline boarder. Walls calcareous, smooth, and finely perforate. Aperture consisting of an interiomarginal, elongate slit, extending from periphery to umbilical boss. Dorsal side (spiral side) slightly convex and evolute. All chambers visible on dorsal side; chambers slowly increasing in size as added and always longer than wide (ratio of 3:1); chambers arranged into four whorls dorsally. Dorsal sutures distinct, narrow, flush, oblique, and curved. Ventral side (umbilical side) strongly convex; umbilicus substituted by well-developed hyaline boss of glassy material. Only chambers of final whorl visible on ventral side, 7–8 in number. Ventral sutures distinct, radiate, narrow, and sigmoid; flush and slightly depressed toward the periphery.

Occurrence. – Maastrichtian and Danian at Sites 525 and 527.

Remarks. – This well-known species is one of the most frequent forms in the sediments analyzed here, and, together with *Gavelinella beccariiformis*, it constitutes the main component in the 'Velasco-type' fauna. A relatively wide species concept was used by Tjalsma & Lohmann (1983), who also included morphotypes such as *N. bronnimanni* (Cushman & Renz) and *N. carinotruempyi* (Finlay) in the concept of *N. truempyi* (Nuttall). A wide species concept of *N. truempyi* is also applied in this study. This species is previously reported from the South Atlantic by Dailey (1983) under the name of *N. truempyi* and by Beckmann (1978) under the name of *N. bronnimanni*. Sliter (1977) figured a specimen, referred by him to *Osangularia cordieriana* (d'Orbigny), which has small chamberlets around the center of the umbilical side and a slit-like aperture, extending from the 'umbilical boss' to the periphery, which suggests that the specimen should be referred to *Nuttallides*.

Genus *Nuttallinella* Belford, 1959

Type species and diagnosis. – See Loeblich & Tappan (1988, p. 603).

Nuttallinella florealis (White, 1928)

Fig. 23A–C

Synonymy. – □*1928b *Gyroidina florealis* n.sp. – White, p. 293, Pl. 10:3. □v1946 *Pulvinulinella? florealis* (White) – Cushman & Renz, p. 46, Pl. 8:4–5. □1977 *Nuttallinella florealis* (White) – Sliter, p. 675, Pl. 6:2–3. □1983 *Nuttallinella florealis* (White) – Dailey, p. 767, Pl. 3:4,8. □1986

Nuttallinella florealis (White) – Van Morkhoven *et al.*, p. 356, Pl. 115. □1991 *Nuttallinella florealis* (White) – Nomura, p. 22, Pl. 2:11. □v1992a *Nuttallinella florealis* (White) – Widmark & Malmgren, p. 112, Pl. 2:4.

Material. – Forty-three specimens.

Description. – Test trochospiral, planoconvex to unequally biconvex, and lenticular in section; subcircular and lobated in outline. Periphery sharp with large irregular hyaline keel. Walls calcareous, smooth, and finely perforate. Aperture consisting of an interiomarginal, elongate slit, extending from periphery to umbilicus. Dorsal side (spiral side) planoconvex to slightly convex; evolute. All chambers visible on dorsal side, about seven chambers in the final whorl. Dorsal sutures limbate, flush, and curved. Ventral side (umbilical side) strongly convex, bell-shaped toward the periphery; involute with small open umbilicus. Only chambers of final whorl visible on ventral side; about seven in number. Ventral sutures indistinct, radial, slightly curved, and depressed.

Occurrence. – Common in the Danian of Site 525, but rare in the Maastrichtian of Site 525 and Danian of Site 527.

Remarks. – This species is easy to recognize because of its bell-shaped umbilical side in side view, strongly planoconvex test shape and large, irregular hyaline keel at the periphery. The examined plesiotype given above and additional secondary types (Cush. Coll. No. 46782; USNM No. 382166), filed under the name of *Pulvinulinella florealis* (White) are slightly different from the specimens found here in being more flaring and concave dorsally.

Nuttallinella sp. A

Fig. 23D–F

Synonymy. – □1983 *Nuttallinella* sp. – Dailey, p. 767, Pl. 3:9–10. □v1988 *Nuttallinella* sp. a – Widmark & Malmgren, p. 71, Pl. 2:2. □v1992a *Nuttallinella* sp. A – Widmark & Malmgren, p. 112, Pl. 2:5. □v1992b *Nuttallinella* sp. A – Widmark & Malmgren, p. 398, Pl. 5:7–9.

Material. – 166 specimens.

Description. – Test trochospiral and strongly planoconvex. Periphery angular; subcircular and lobated in outline. Walls calcareous and finely perforate. Aperture consisting of an interiomarginal, elongate slit, extending from periphery to umbilicus. Dorsal side (spiral side) planoconvex to slightly convex; evolute. All chambers visible on dorsal side, slowly increasing in size as added; about eight in number in the final whorl. Dorsal sutures distinct, flush, and curved. Ventral side (umbilical side) strongly convex with small almost closed umbilicus; involute. Only chambers of final whorl visible on ventral side;

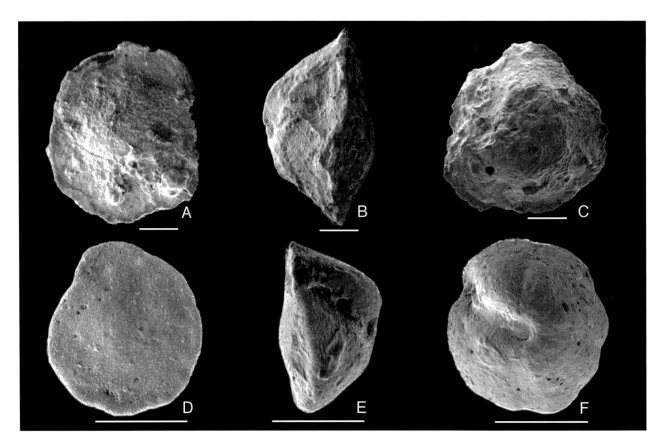

Fig. 23. Scale 100 μm. □A–C. *Nuttallinella florealis* (White); sample 525A-40-1, 140–141 cm (Danian). □A. Spiral view. □B. Peripheral view. □C. Umbilical view. □D–F. *Nuttallinella* sp. A; sample 527-32-5, 57–58 cm (Maastrichtian). □D. Spiral view. □E. Peripheral view. □F. Umbilical view.

about eight in number. Ventral sutures distinct, slightly depressed, radial, and slightly curved.

Occurrence. – Danian and Maastrichtian at Sites 525 and 527.

Remarks. – This is a minute form with uncertain taxonomic status. It resembles the specimen of *Nuttallinella* sp. reported by Dailey (1983) from the Rio Grande Rise.

Superfamily Nonionacea Schultze, 1854

Diagnosis. – See Loeblich & Tappan (1988, p. 615).

Family Nonionidae Schultze, 1854

Diagnosis. – See Loeblich & Tappan (1988, p. 615).

Subfamily Nonioninae Schultze, 1854

Diagnosis. – See Loeblich & Tappan (1988, p. 615).

Genus *Nonion* de Montfort, 1808

Type species and diagnosis. – See Loeblich & Tappan (1988, p. 617).

Nonion spp.

Fig. 24A–B

Material. – Forty-six specimens.

Occurrence. – Danian and Maastrichtian at Sites 525 and 527.

Remarks. – Several small, planispiral, involute, and compressed specimens with rounded periphery were assigned to this taxon.

Genus *Nonionella* Cushman, 1926

Type species and diagnosis. – See Loeblich & Tappan (1988, p. 617).

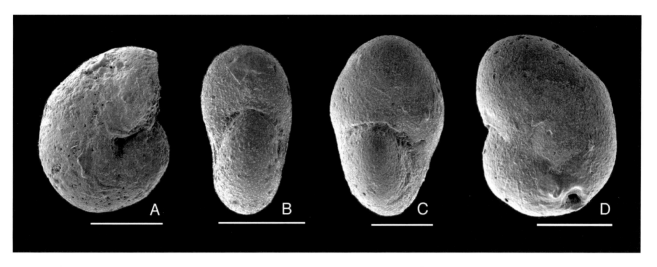

Fig. 24. Scale 100 μm. □A–B. *Nonion* spp.; sample 525A-40-2, 69–70 cm (Maastrichtian). □A. Side view. □B. Peripheral view. □C–D. *Nonionella* spp.; sample 525A-40-2, 59–60 cm (Maastrichtian). □C. Spiral view. □D. Peripheral view.

Nonionella spp.

Fig. 24C–D

Material. – Twenty-three specimens.

Occurrence. – Danian and Maastrichtian at Sites 525 and 527.

Remarks. – A few, poorly preserved, specimens were assigned to this taxon. They are characterized by having small, trochospiral, and compressed tests with a more or less evolute spiral side, involute umbilical side, and a rounded periphery.

Subfamily Pulleniinae Schwager, 1877

Diagnosis. – See Loeblich & Tappan (1988, p. 620).

Genus *Pullenia* Parker & Jones, 1862

Type species and diagnosis. – See Loeblich & Tappan (1988, p. 621).

Pullenia cf. *cretacea* Cushman, 1936

Fig. 25A–B

Synonymy. – □vcf. *1936 *Pullenia cretacea* n.sp. – Cushman, p. 75, Pl. 13:8. □v1992a *Pullenia* cf. *cretacea* Cushman – Widmark & Malmgren, p. 12, Pl. 9:5.

Material. – 109 specimens.

Description. – Test planispiral, biumbilicate, and involute; oval in apertural view. Periphery broadly rounded. Seven chambers in final whorl; gradually increasing in size as added. Sutures distinct, narrow, nearly straight, and slightly depressed. Walls calcareous, perforate, and smooth. Aperture consisting of an elongate slit at base of final chamber, extending from one umbilicus to the other.

Occurrence. – Danian and two specimens in the youngest Maastrichtian sample at Site 527.

Remarks. – *Pullenia* cf. *cretacea* is intermediate between *P. coryelli* and *P.* sp. compressed form with respect to the sphericity in apertural view. The specimens referred to this taxon are very similar to the holotype of *Pullenia cretacea* Cushman and the paratypes described by Cushman (1936). The specimens differ, however, in having 6–7 chambers in the final whorl, instead of 5 chambers as in the holotype. Dailey (1983) figured a specimen under the name of *P. cretacea* Cushman, which probably is referable to *Pullenia reussi* Cushman & Todd. The holotype of *P. reussi*, which was described by Cushman & Todd (1943), also has a low, equatorially flattened apertural face and 4½ chambers in the final whorl, as in the specimen figured by Dailey (1983).

Pullenia coryelli White, 1929

Fig. 25C–D

Synonymy. – □*1929 *Pullenia coryelli* n.sp. – White, p. 56, Pl. 5:22. □v1944 *Pullenia coryelli* White – Cushman & Deaderick, p. 339, Pl. 53:26. □v1946 *Pullenia coryelli* White – Cushman & Renz, p. 47, Pl. 8:9. □1977 *Pullenia*

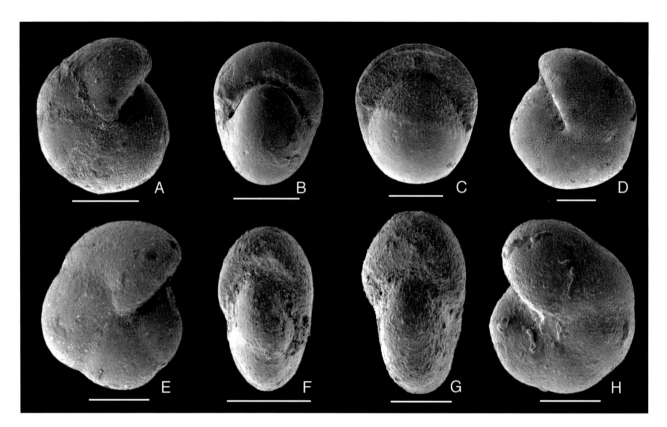

Fig. 25. Scale 100 μm. □A–B. *Pullenia* cf. *cretacea* Cushman; sample 527-32-3, 139–140 cm (Danian). □A. Side view. □B. Apertural view. □C–D. *Pullenia coryelli* White; sample 525A-40-2, 69–70 cm (Maastrichtian). □C. Apertural view. □D. Side view. □E–F. *Pullenia jarvisi* Cushman. □E. Side view; sample 525A-40-2, 78–79 cm (Maastrichtian). □F. Apertural view; sample 525A-40-2, 69–70 cm (Maastrichtian). □G–H. *Pullenia* sp. compressed form; sample 527-32-4, 39–40 cm (Maastrichtian). □G. Apertural view. □H. Side view.

coryelli White – Sliter, p. 675, Pl. 8:5. □1978 *Pullenia coryelli* White – Beckmann, p. 768, Pl. 3:21. □1983 *Pullenia coryelli* White – Dailey, p. 768, Pl. 5:1–2. □1983 *Pullenia coryelli* White – Tjalsma & Lohmann, p. 18, Pl. 5:5. □1990 *Pullenia coryelli* White – Thomas, p. 590, Pl. 3:6. □1991 *Pullenia coryelli* White – Nomura, p. 22, Pl. 4:9. □v1992a *Pullenia coryelli* White – Widmark & Malmgren, p. 112, Pl. 9:4.

Material. – Fifty-eight specimens.

Description. – Test planispiral, completely involute, and subglobular; biumbonate. Periphery broadly rounded. Six to seven chambers in the final whorl, slowly increasing in size as added; last chamber with low apertural face. Sutures distinct, straight or weakly curved, and very slightly depressed. Walls calcareous, finely perforate, and smooth. Aperture consisting of an elongate slit at base of final chamber, extending from one umbo to the other.

Occurrence. – Danian and Maastrichtian at Sites 525 and 527.

Remarks. – This well-known species is easy to recognize because of its globular test shape.

Pullenia jarvisi Cushman, 1936

Fig. 25E–F

Synonymy. – □v*1936 *Pullenia jarvisi* n.sp. – Cushman, p. 77, Pl. 13:6. □1983 *Pullenia jarvisi* Cushman – Dailey, p. 768, Pl. 5:3–4. □v1988 *Pullenia quinqueloba* (Reuss) – Widmark & Malmgren, p. 52. □1992 *Pullenia jarvisi* Cushman – Gawor-Biedowa, p. 146, Pl. 30:3–4.

Material. – Eleven specimens.

Description. – Test planispiral, moderately compressed, completely involute, subcircular and lobated in outline; biumbilicate. Periphery rounded. About five inflated chambers in final whorl, increasing rapidly in size as added; final chamber with large apertural face. Sutures distinct, curved, and depressed. Walls calcareous, finely perforate, and smooth. Aperture consisting of an elongate slit at base of final chamber, extending from one umbilicus to the other.

Occurrence. – Maastrichtian at Site 525, Danian and Maastrichtian at Site 527.

Remarks. – Specimens encountered in the material from Site 527 were referred to *Pullenia quinqueloba* (Reuss) by Widmark & Malmgren (1988) but are here considered as belonging to *Pullenia jarvisi* Cushman. *Pullenia jarvisi* is here separated from other *Pullenia* on the basis of its relatively compressed test, lobated outline, and large apertural face. Comparison between the holotype given above and the specimens here referred to this species made this identification reliable.

Pullenia sp. compressed form

Fig. 25G–H

Material. – 139 specimens.

Description. – Test planispiral, biumbilicate, and involute. Periphery rounded. Six chambers in final whorl; rapidly increasing in size as added; last chamber inflated. Sutures distinct, narrow, curved, and depressed. Walls calcareous, perforate, and smooth. Aperture consisting of an elongate slit at base of final chamber, extending from one umbilicus to the other; large apertural face.

Occurrence. – Danian and Maastrichtian at Sites 525 and 527.

Remarks. – Specimens with an inflated last chamber, compressed test in apertural view, and rapidly increasing chambers when added are included in this taxon.

Superfamily Chilostomellacea Brady, 1881

Diagnosis. – See Loeblich & Tappan (1988, p. 624).

Family Chilostomellidae Brady, 1881

Diagnosis. – See Loeblich & Tappan (1988, p. 624).

Subfamily Chilostomellinae Brady, 1881

Diagnosis. – See Loeblich & Tappan (1988, p. 624).

Genus *Allomorphina* Reuss *in* Czjzek, 1849

Type species and diagnosis. – See Loeblich & Tappan (1988, p. 624).

Allomorphina cf. *navarroana* Cushman, 1936

Fig. 26A–B

Synonymy. □v*cf. 1936 *Allomorphina navarroana* n.sp. – Cushman, p. 73, Pl. 13:1.

Material. – Two specimens.

Description. – Test trochospiral, small, and biconvex; nearly oval in outline. Periphery rounded. Aperture consisting of an elongate slit at base of last chamber. Walls calcareous, smooth, and perforate. Last chamber strongly overlaps the previous ones and makes up about ¾ of total length of test. Sutures indistinct and hardly visible. Dorsal side (spiral side) convex with depressed spire. Ventral side (umbilical side) convex with large umbilical flap.

Occurrence. – Maastrichtian at Site 525.

Remarks. – Two specimens of this distinct form were encountered in one sample only. Their generally oval test shape resembles that of the examined holotype of *Allomorphina navarroana* Cushman (1936), but they differ from the holotype in being more compressed.

Allomorphina minuta Cushman, 1936

Fig. 26D–E

Synonymy. – □v*1936 *Allomorphina minuta* n.sp. – Cushman, p. 72, Pl. 13:3. □1983 *Allomorphina minuta* Cushman – Dailey, p. 764, Pl. 4:7. □1992 *Quadrimorphina minuta* Cushman – Gawor-Biedowa, p. 147, Pl. 30:6.

Material. – Six specimens.

Description. – Test trochospiral, small, planoconvex to concavoconvex; subtriangular in outline. Periphery rounded. Aperture hardly visible in the few specimens encountered, but seems to consist of a basal elongate slit at final chamber. Walls calcareous, smooth, and perforate both dorsally and ventrally. Dorsal side (spiral side) slightly convex with depressed spire; evolute. Three inflated, overlapping chambers per whorl dorsally, rapidly increasing in size as added. Dorsal sutures indistinct (except in the last whorl), narrow, and slightly depressed. Ventral side (umbilical side) nearly flat and involute; umbilicus covered by umbilical flap. Only chambers of final whorl visible ventrally; three inflated chambers in final whorl. Ventral sutures fairly distinct, straight, and slightly depressed.

Occurrence. – Maastrichtian at Sites 525 and 527.

Remarks. – This scarce and small species is distinguished by its small test with three inflated chambers per whorl, which gives the test its subtriangular outline. Specimens

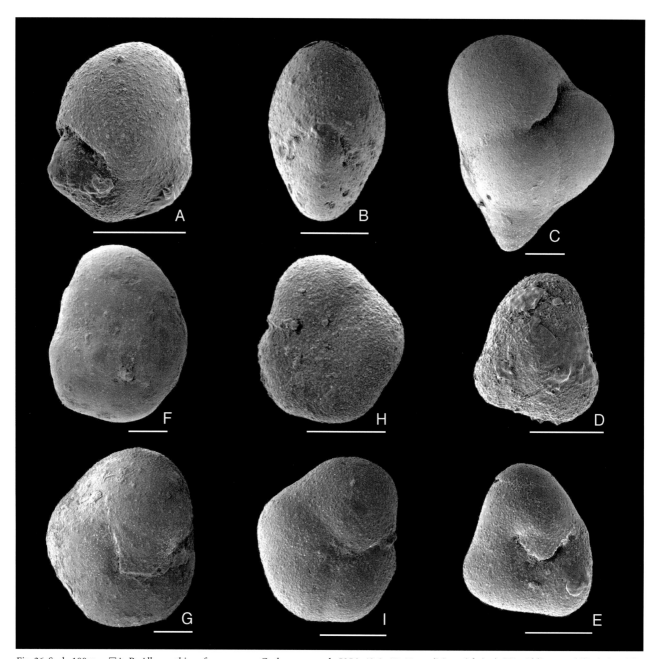

Fig. 26. Scale 100 µm. ☐A–B. *Allomorphina* cf. *navarroana* Cushman; sample 525A-40-2, 47–48 cm (Maastrichtian). ☐A. Oblique umbilical view. ☐B. Peripheral view. ☐C. *Allomorphina trochoides* (Reuss); oblique umbilical view; sample 527-32-4, 87–89 cm (Maastrichtian). ☐D–E. *Allomorphina minuta* Cushman. ☐D. Spiral view; sample 527-32-4, 57–59 cm (Maastrichtian); ☐E. Umbilical view; sample 527-32-4, 107–108 cm (Maastrichtian). ☐F–G. *Quadrimorphina allomorphinoides* (Reuss). ☐F. Spiral view; sample 527-32-4, 29–30 cm (Danian). ☐G. Umbilical view; sample 527-32-4, 39–40 cm (Danian). ☐H–I. *Quadrimorphina camerata* (Brotzen) sample 527-32-4, 29–30 cm (Danian). ☐H. Spiral view. ☐I. Umbilical view.

here referred to this species were found to be consistent with its holotype.

Allomorphina trochoides (Reuss, 1845)

Fig. 26C

Synonymy. – ☐*1845 *Globigerina trochoides* n.sp. – Reuss, p. 36, Pl. 12:22. ☐1932 *Allomorphina trochoides* (Reuss) – Cushman & Jarvis, p. 49, Pl. 15:3. ☐1977 *Allomorphina trochoides* (Reuss) – Sliter, p. 674, Pl. 8:2. ☐1983 *Allomorphina trochoides* (Reuss) – Dailey, p. 766, Pl. 4:8. ☐1991 *Praebulimina* sp. B – Nomura, Pl. 1:21. ☐1992 *Allomorphina trochoides* (Reuss) – Gawor-Biedowa, p. 147, Pl. 27:14.

Material. – Four specimens.

Description. – Test trochospiral, cone-shaped in side view, and subtriangular in transection. Periphery rounded. Aperture consisting of an elongate slit at base of last chamber. Walls calcareous, smooth, and perforate. Dorsal side (spiral side) cone-shaped with a pointing initial end; evolute. Dorsal chambers triserially arranged and hardly visible in early cone-shaped portion; last whorl consisting of three inflated chambers. Dorsal sutures indistinct in early portion; distinct and depressed in later portion. Ventral side (umbilical side) slightly convex and involute. Only chambers of last whorl visible ventrally; chambers strongly inflated; last chamber strongly overlaps the previous ones. Ventral sutures distinct, straight, and depressed.

Occurrence. – Danian and Maastrichtian at Site 527.

Remarks. – This species is very easy to recognize on the basis of its curious cone-like and triserial test shape.

Family Quadrimorphinidae Saidova, 1981

Diagnosis. – See Loeblich & Tappan (1988, p. 627).

Genus *Quadrimorphina* Finlay, 1939

Type species and diagnosis. – See Loeblich & Tappan (1988, p. 627).

Quadrimorphina allomorphinoides (Reuss),1860

Fig. 26F–G

Synonymy. – □*1860 *Valvulina allomorphinoides* nov. sp. – Reuss, p. 223, Pl. 11:6. □*v*1931a *Valvulineria allomorphinoides* (Reuss) – Cushman, p. 53, Pl. 9:6. □*v*1931b *Valvulineria allomorphinoides* (Reuss) – Cushman, p. 43, Pl. 6:2. □1977 *Quadrimorphina allomorphinoides* (Reuss) – Sliter, p. 675, Pl. 8:1. □1978 *Valvulineria allomorphinoides* (Reuss) – Beckmann, p. 769, Pl. 2:25–27. □1983 *Quadrimorphina allomorphinoides* (Reuss) – Dailey, p. 768, Pl. 4:9. □1991 *Quadrimorphina allomorphinoides* (Reuss) – Nomura, p. 23, Pl. 1:23.

Material. – 136 specimens.

Description. – Test trochospiral, biconvex, and ovoid in outline. Periphery rounded. Walls calcareous, finely perforate, and smooth. Aperture consisting of an interiomarginal slit, extending from periphery to umbilicus. Dorsal side (spiral side) convex, not completely involute, leaving the spiral suture visible throughout. Only chambers of

final whorl visible in full on dorsal side; about 3–4 chambers in final whorl; last chamber strongly inflated, making up about ½ of total length of test; chambers increase gradually in size as added. Dorsal sutures distinct, narrow, slightly curved, and slightly depressed. Ventral side (umbilical side) convex and involute; umbilicus covered by a large flap (= extension of last chamber), which sometimes is broken away. Only chambers of last whorl visible on ventral side. Ventral sutures distinct, narrow, straight, and depressed.

Occurrence. – Danian and Maastrichtian at Sites 525 and 527.

Remarks. – This very distinctive species is separated from *Q. camerata* on the basis of its elongated and inflated final chamber and the more rapid chamber enlargement on the dorsal side. The plesiotypes given above are consistent with the concept of *Quadrimorphina allomorphinoides* (Reuss) applied here, in having a strongly enlarged last chamber and about four chambers in the final whorl.

Quadrimorphina camerata (Brotzen, 1936)

Fig. 26H–I

Synonymy. – □*v*1932 *Valvulineria allomorphinoides* (Reuss) – Cushman & Jarvis, p. 46, Pl. 13:17. □*1936 *Valvulineria camerata* n.sp. – Brotzen, p. 155, Pl. 10:2; Text-fig. 57 (1–2). □*v*1941 *Valvulineria allomorphinoides* (Reuss) – Cushman & Hedberg, p. 96, Pl. 23:9. □*v*1946 *Valvulineria allomorphinoides* (Reuss) – Cushman & Renz, p. 44, Pl. 7:13–14. □1978 *Valvulineria camerata* Brotzen – Beckmann, p. 769, Pl. 2:30–31. □*v*1992b *Quadrimorphina camerata* (Brotzen) – Widmark & Malmgren, p. 402, Pl. 6:4–6.

Material. – Twenty-four specimens.

Description. – Test trochospiral, weakly biconvex, and ovoid to subcircular in outline. Periphery rounded. Walls calcareous, perforate, and smooth. Aperture consisting of an interiomarginal slit, extending from periphery to umbilicus. Dorsal side (spiral side) weakly convex and evolute. All chambers visible on dorsal side; about five chambers in final whorl; last chamber somewhat inflated, making up about ⅓ of total length of test; chambers increasing gradually in size as added. Dorsal sutures distinct, narrow, curved, and slightly depressed. Ventral side (umbilical side) convex and involute; umbilicus covered by a large flap (= extension of last chamber), which sometimes is broken away. Only chambers of final whorl visible on ventral side. Ventral sutures distinct, narrow, straight, and depressed.

Occurrence. – Maastrichtian at Site 525, Danian and Maastrichtian at Site 527.

Remarks. – *Quadrimorphina camerata* is separated from *Q. allomorphinoides* by its smaller final chamber (comprising about $\frac{1}{3}$ of total length of test instead of $\frac{1}{2}$), its more open evolute spiral side, greater number of chambers in the final whorl, and more gradual chamber enlargement dorsally. The plesiotypes given above are considered to be within the concept of *Q. camerata* rather than that of *Q. allomorphinoides* on the basis of their greater number of chambers in the final whorl (5–7) and their less inflated final chamber. About 20 specimens of *Valvulineria camerata* Brotzen, collected from the type locality of this species (Eriksdal, Sweden) and identified by Brotzen, were available and examined at the Naturhistoriska Riksmuseet, Stockholm. These specimens agreed fairly well with the specimens in the present material. They all have about five chambers in the final whorl and a last chamber that makes up about $\frac{1}{3}$ of the total test length. Furthermore, Brotzen's specimens and those in the present material are also found to be more compressed dorsally–ventrally than the specimens of *Q. allomorphinoides* found here.

Family Alabaminidae Hofker, 1951

Diagnosis. – See Loeblich & Tappan (1988, p. 627).

Genus *Alabamina* Toulmin, 1941

Type species and diagnosis. – See Loeblich & Tappan (1988, p. 627).

Alabamina sp. A

Fig. 27A–C

Synonymy. – □v1970 *Eponides* sp. A – Todd, p. 145, Pl. 3:1. □1983 *Alabamina creta* (Finlay) – Dailey, p. 764, Pl. 5:5–7. □v1988 *Alabamina* sp. a – Widmark & Malmgren, p. 73, Pl. 3:1. □v1992a *Alabamina creta* (Finlay) – Widmark & Malmgren, p. 73, Pl. 3:1.

Material. – 109 specimens.

Description. – Test trochospiral, unequally biconvex, and subcircular in outline. Periphery rounded to subangular. Aperture basal, extending from an invagination in apertural face at final chamber to halfway toward umbilical area. Walls calcareous, finely perforate, and smooth both dorsally and ventrally. Dorsal side (spiral side) weakly convex and evolute; all dorsal chambers visible, about six weakly lobated chambers in final whorl. Dorsal sutures

distinct, narrow, curved, and slightly depressed. Ventral side (umbilical side) convex and involute; umbilicus closed. Only chambers of last whorl visible ventrally; about six chambers in final whorl. Ventral sutures distinct, curved, and depressed.

Occurrence. – Maastrichtian at Site 525, Danian and Maastrichtian at Site 527.

Remarks. – The specimens referred to this species are very close to the form named *Eponides* sp. A by Todd (1970) and show a close resemblance to the figures of *A. creta* (Finlay) given by Dailey (1983). Widmark & Malmgren (1992a) referred erroneously this form to *A. creta* (Finlay), which has nearly straight dorsal sutures that are tangential to the spiral suture (see, e.g., Tjalsma & Lohmann 1983, Pl. 7:13c); the present form has more or less orthogonal dorsal sutures.

Alabamina sp. C

Fig. 27D–F

Material. – Ten specimens.

Description. – Test trochospiral and biconvex; subcircular and strongly lobated in outline. Periphery acute. Aperture basal, extending from an invagination in apertural face at final chamber to halfway toward umbilical area. Wall calcareous and finely perforate both dorsally and ventrally. Dorsal side (spiral side) weakly convex and evolute. All chambers visible dorsally; about six lobated chambers in final whorl. Dorsal sutures distinct, curved, and slightly depressed. Ventral side (umbilical side) convex and evolute. Only chambers of final whorl visible ventrally. Ventral sutures distinct, sinuate, and depressed.

Occurrence. – Danian and Maastrichtian at Sites 525 and 527.

Remarks. – The taxonomic status of this form is uncertain owing to its scarcity in the material. *Alabamina* sp. C differs, however, from *Alabamina* sp. A by its acute periphery and strongly lobated outline.

Genus *Valvalabamina* Reiss, 1963

Type species and diagnosis. – See Loeblich & Tappan (1988, p. 628).

Valvalabamina sp. evolute form

Fig. 28A–C

Synonymy. – □1983 *Gyroidinoides depressus* (Alth) – Dailey, p. 767, Pl. 6:10–12. □1983 *Gyroidinoides* sp., Tjalsma

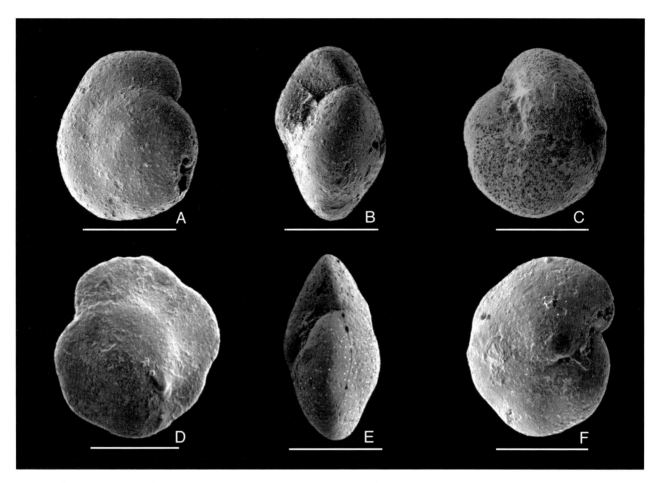

Fig. 27. Scale 100 μm. □A–C. *Alabamina* sp. A; sample 525A-40-3, 70–71 cm (Maastrichtian). □A. Spiral view. □B. Peripheral view. □C. Umbilical view. □D–F. *Alabamina* sp. C. □D. Spiral view; sample 525A-40-2, 130–131 cm (Maastrichtian). □E. Peripheral view; sample 525A-40-2, 92–93 cm (Maastrichtian). □F. Umbilical view; sample 525A-40-2, 92–93 cm (Maastrichtian).

& Lohmann, p. 15, Pl. 7:7. □v1992a *Valvalabamina* sp. evolute form – Widmark & Malmgren, p. 113, Pl. 3:3. □v1992b *Valvalabamina* sp. evolute form – Widmark & Malmgren, p. 402, Pl. 3:10–12.

Material. – Ninety-four specimens.

Description. – Test trochospiral, planoconvex to unequally biconvex; subcircular in outline. Periphery rounded. Walls calcareous and finely perforate. Aperture consisting of an interiomarginal slit along apertural face. Dorsal side (spiral side) flat to weakly convex and evolute. All chambers visible on dorsal side; about eight chambers in final whorl. Dorsal sutures distinct, strongly (often abruptly) curved, and slightly depressed. Ventral side (umbilical side) moderately convex and involute; umbilicus open. Only chambers of final whorl visible on ventral side. Ventral sutures distinct, narrow, slightly curved, and slightly depressed.

Occurrence. – Danian and Maastrichtian at Sites 525 and 527.

Remarks. – Specimens referred to this taxon are close to the specimens illustrated by Dailey (1983) as *Gyroidinoides depressus* (Alth) and by Tjalsma & Lohmann (1983) as *Gyroidinoides* sp. The generic status of their assignment to *Gyroidinoides* is, however, in doubt because of the low convexity of the ventral side in this form. It resembles *Valvalabamina* sp. involute form in most respects, but is different in having an evolute spiral side. Several secondary types at the Smithsonian Institution were examined and compared with the specimens here referred to this taxon. Two plesiotypes, named *Gyroidina depressa* (Alth), and figured by Cushman (1944a, b), were different from the specimens in the present material in being more lobated in outline and in having more distinct, narrow, and depressed sutures. Two topotypes of *Rotalia depressa* Alth from the Cretaceous 'Mueronata Kreide' (Lemberg, Galecia, Germany) were also examined. The topotypes were found to be different from the plesiotypes examined in having an unlobated outline and very indistinct sutures both dorsally and ventrally, which

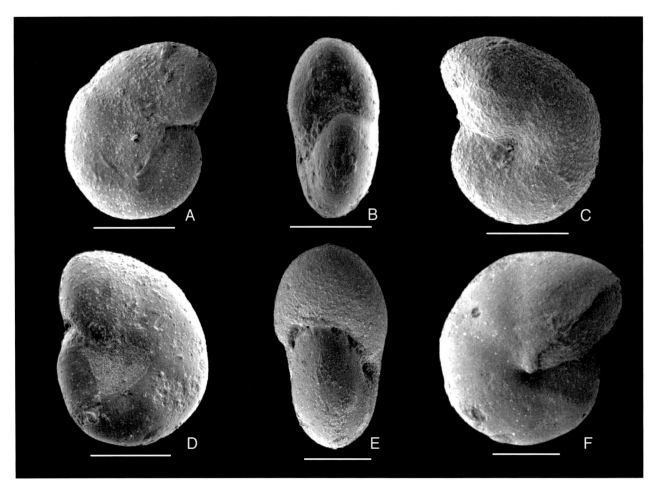

Fig. 28. Scale 100 µm. □A–C. *Valvalabamina* sp. evolute form; sample 527-32-4, 0–1 cm (Danian). □A. Spiral view. □B. Peripheral view. □C. Umbilical view. □D–F. *Valvalabamina* sp. involute form; sample 527-32-4, 0–1 cm (Danian). □D. Spiral view. □E. Peripheral view. □F. Umbilical view.

demonstrates the inconsistency in the use of the concept of Alth's species.

Valvalabamina sp. involute form

Fig. 28D–F

Synonymy. – □v1970 *Gavelinella* sp., Todd, p. 147, Pl. 4:7. □v1988 *Anomalina* sp. a – Widmark & Malmgren, p. 75, Pl. 3:3. □v1992a *Valvalabamina* sp. involute form – Widmark & Malmgren, p. 113, Pl. 3:2. □v1992b *Valvalabamina* sp. involute form – Widmark & Malmgren, p. 402, Pl. 2:4–6.

Material. – 236 specimens.

Description. – Test trochospiral and unequally biconvex. Periphery rounded. Aperture consisting of an interiomarginal slit, extending from periphery to umbilicus. Walls calcareous, distinctly perforate, and smooth. Dorsal side (spiral side) weakly convex to weakly concave, depending on the degree of involution of the spiral side; not completely involute. Only chambers of final whorl totally visible dorsally; chambers of inner whorls only partly visible; about seven chambers in final whorl; chambers gradually increasing in size as added. Sutures distinct, narrow, curved, and depressed. Ventral side (umbilical side) convex (more than dorsally) and involute; umbilicus small and open. Only chambers of final whorl visible ventrally. Ventral sutures distinct, nearly straight, and depressed.

Occurrence. – Danian and Maastrichtian at Sites 525 and 527.

Remarks. – The generic status of this taxon is uncertain. It resembles *Valvalabamina* sp. evolute form in most respects, but is different in having a more or less involute spiral side, which is evolute in *Valvalabamina* sp. evolute form. *Valvalabamina* sp. involute form differs from the nearly planispiral *V.? praeacuta* in being trochospiral and in having narrow sutures, which are nearly straight on the ventral side instead of broad and curved as in *V.? praea-*

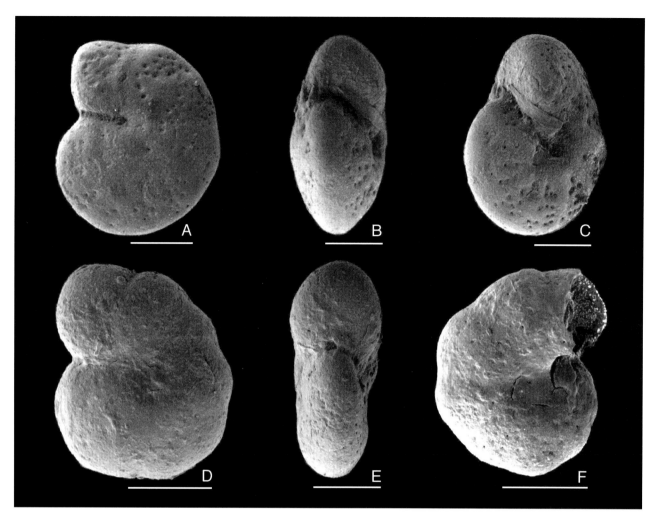

Fig. 29. Scale 100 µm. □A–C. *Valvalabamina? praeacuta* (Vasilenko); sample 527-32-4, 19–20 cm (Danian). □A. Spiral view. □B. Peripheral view. □C. Oblique umbilical view. □D–F. *Valvalabamina?* sp. lobated form; sample 527-32-4, 19–20 cm (Danian). □D. Spiral view. □E. Peripheral view. □F. Umbilical view.

cuta. Examined specimens referred to as *Gavelinella* sp. by Todd (1970) were found to be consistent with the concept of this taxon used here.

Valvalabamina? praeacuta (Vasilenko, 1950)

Fig. 29A–C

Synonymy. – □*1950 *Anomalina praeacuta* n.sp. – Vasilenko, p. 208, Pl. 5:2–3. □1978 *Anomalina praeacuta* Vasilenko – Proto Decima & Bolli, Pl. 5:9–11. □1983 *Anomalina praeacuta* Vasilenko – Tjalsma & Lohmann, p. 4, Pl. 4:10; Pl. 7:8. □v1988 *Anomalina praeacuta* Vasilenko – Widmark & Malmgren, p. 75, Pl. 3:2. □v[non] 1994 *Anomalinoides praeacuta* (Vasilenko) – Speijer, p. 60, Pl. 8:1.

Material. – Seventy-four specimens.

Description. – Test nearly planispiral, biconvex, and compressed; subcircular in outline. Periphery subacute. Aperture consisting of an interiomarginal slit extending to umbilicus. Dorsal side (spiral side) slightly convex and evolute. All chambers visible dorsally; 8–9 in number. Dorsal sutures distinct, broad, curved, and flush. Walls calcareous with relatively large pores scattered all over the test; umbilicus surrounded and partly closed by thick, hyaline material. Ventral side (umbilical side) slightly convex and involute. Only chambers of final whorl visible ventrally. Ventral sutures distinct, broad, curved, and flush.

Occurrence. – Danian and Maastrichtian at Sites 525 and 527.

Remarks. – This species is distinguished from other similar forms in having a nearly planispiral test and evolute dorsal side and by the characteristic thickening at the umbilical area. Since a petition applying for suppression of the genus *Anomalina* have been sent to the ICZN (Loeblich & Tappan 1988; p. 605), this species is here tentatively referred to *Valvalabamina?*.

Valvalabamina? sp. lobated form

Fig. 29D–F

Material. – Seven specimens.

Description. – Test weakly trochospiral (nearly planispiral) and lobated in outline. Periphery rounded. Aperture consisting of an interiomarginal slit, extending from periphery to umbilicus. Walls calcareous, perforate, and smooth. Dorsal side (spiral side) flat and not completely involute. Only chambers of final whorl completely visible dorsally; chambers of inner whorls only partly visible; about nine inflated chambers in final whorl; chambers increasing rapidly in size as added. Dorsal sutures distinct, curved, and distinctly depressed. Ventral side (umbilical side) flat to weakly concave and involute; umbilicus open. Only chambers of final whorl visible ventrally. Ventral sutures distinct, slightly curved, and depressed.

Occurrence. – Danian and Maastrichtian at Site 525.

Remarks. – The generic status of this rare form is uncertain. Only a few specimens of the taxon were encountered in a few samples. *Valvalabamina* sp. lobated form is separated from other *Valvalabamina* by its lobated outline and depressed sutures.

Family Globorotalitidae Loeblich & Tappan, 1984

Diagnosis. – See Loeblich & Tappan (1988, p. 628).

Genus *Globorotalites* Brotzen, 1942

Type species and diagnosis. – See Loeblich & Tappan (1988, p. 629).

Globorotalites cf. *conicus* (Carsey, 1926)

Fig. 30A–C

Synonymy. – □cf. 1926 *Truncatulina refulgens* Montfort var. *conica* Carsey, – Carsey, p. 46, Pl. 4:15. □vcf. 1956 *Globorotalites conicus* (Carsey) – Said & Kenawy, p. 147, Pl. 4:43. □1978 *Globorotalites conicus* (Carsey) – Beck-

mann, p. 766, Pl. 4:22–23. □1983 *Globorotalites conicus* (Carsey) – Dailey, p. 767, Pl. 5:9–10. □1990 *Globorotalites conicus* (Carsey) – Thomas, p. 590, Pl. 2:1–2. □1991 *Globorotalites conicus* (Carsey) – Nomura, p. 22, Pl. 5:2.

Material. – Twenty-one specimens.

Description. – Test trochospiral, planoconvex, and subcircular in outline. Periphery acute with narrow hyaline border. Aperture consisting of an interiomarginal slit, extending from periphery to umbilicus. Walls calcareous, finely perforate and smooth. Dorsal side (spiral side) flat to weakly concave, with raised central area (proloculus); evolute. All chambers visible from dorsal side; about six chambers in final whorl. Dorsal sutures distinct, straight, oblique, and flush. Ventral side (umbilical side) strongly convex (nearly cone-shaped), with small open umbilicus; involute. Only chambers of final whorl visible on ventral side. Ventral sutures distinct, narrow, slightly curved, and flush or weakly depressed.

Occurrence. – Maastrichtian at Site 525.

Remarks. – This species is distinguished from other *Globorotalites* encountered here by its small, but distinct, open umbilicus and even, unlobated outline. The hypotype of this species figured by Said & Kenawy (1956) was examined, and it was found to be different from the specimens in the present material in having fewer chambers in the last whorl, lacking the hyaline boarder, and in being less cone-shaped.

Globorotalites sp. A

Fig. 30D–F

Material. – Four specimens.

Description. – Test trochospiral and planoconvex; lobated outline. Periphery angular without hyaline keel or border. Aperture consisting of an interiomarginal slit, extending from periphery to umbilicus; large apertural face. Walls calcareous, perforate, and smooth. Dorsal side (spiral side) flat and evolute. All chambers visible on dorsal side; six chambers in final whorl, rapidly increasing in size as added. Dorsal sutures distinct, oblique, curved, and slightly depressed. Ventral side (umbilical side) strongly convex and involute. Only chambers of final whorl visible on ventral side. Ventral sutures distinct, narrow, sinuate, and depressed.

Occurrence. – Maastrichtian at Site 527.

Remarks. – This distinct species occurs as a few specimens only. It is separated from other *Globorotalites* on the basis of its sinuate ventral sutures, large apertural face, and chambers that increase rapidly as added.

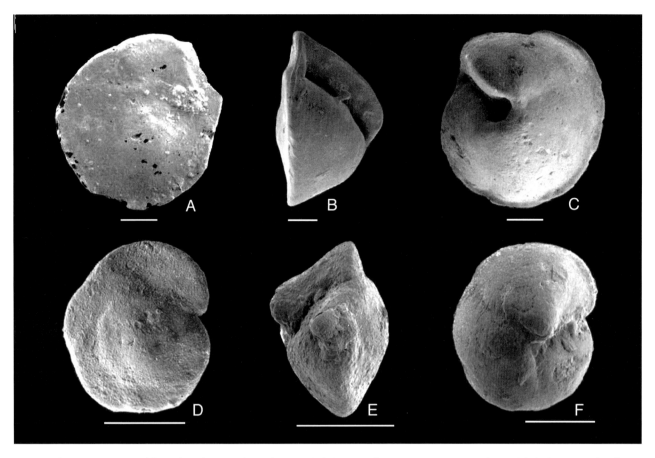

Fig. 30. Scale 100 μm. □A–C. *Globorotalites* cf. *conicus* (Carsey). □A. Spiral view; sample 525A-40-2, 100–102 cm (Maastrichtian). □B. Peripheral view; sample 525A-40-2, 128–129 cm (Maastrichtian). □C. Umbilical view; sample 525A-40-2, 109–110 cm (Maastrichtian). □D–F. *Globorotalites* sp. A. □D. Spiral view; sample 527-32-3, 117–118 cm (Danian). □E. Peripheral view; sample 527-32-4, 29–30 cm (Danian). □F. Umbilical view; sample 527-32-3, 129–130 cm (Danian).

Globorotalites sp. B

Fig. 31A–C

Synonymy. – □v1992a *Globorotalites* sp. B – Widmark & Malmgren, p. 111, Pl. 6:1. □v1992b *Globorotalites* sp. B – Widmark & Malmgren, p. 394, Pl. 3:1–3.

Material. – Sixty-five specimens.

Description. – Test trochospiral, planoconvex with a lobated, subcircular outline. Periphery acute with a narrow hyaline keel. Aperture consisting of an interiomarginal slit with an overhanging lip, extending from periphery to umbilicus. Walls calcareous, perforate, and smooth. Dorsal side (spiral side) flat and evolute. All chambers visible on dorsal side; about seven chambers in last whorl, gradually increasing in size as added. Dorsal sutures distinct, oblique, curved, and slightly depressed. Ventral side (umbilical side) strongly convex, bell-shaped, and involute; umbilicus small and closed. Only chambers of last whorl visible on ventral side. Ventral sutures distinct, slightly curved, and depressed.

Occurrence. – Danian and Maastrichtian at Site 257.

Remarks. – This species is similar to *Globorotalites* sp. C, but it differs by its much smaller test size, more depressed ventral sutures, and in being more convex and cone-shaped ventrally.

Globorotalites sp. C

Fig. 31D–F

Synonymy. – □v1970 *Globorotalites multiseptus* Brotzen – Todd, p. 146, Pl. 3:4.

Material. – Thirty-seven specimens.

Description. – Test trochospiral and planoconvex, with a lobated, subcircular outline. Periphery acute with narrow hyaline keel. Aperture consisting of interiomarginal slit with overhanging lip, extending from periphery to umbilicus. Walls calcareous, perforate, and smooth. Dorsal side (spiral side) flat to weakly convex and evolute. All

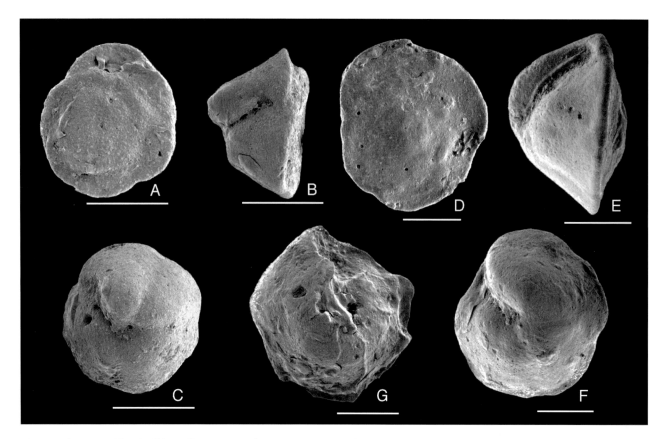

Fig. 31. Scale 100 µm. □A–C. *Globorotalites* sp. B; sample 527-32-5, 92–94 cm (Maastrichtian). □A. Spiral view. □B. Peripheral view. □C. Umbilical view. □D–F. *Globorotalites* sp. C; sample 527-32-5, 33–34 cm (Maastrichtian). □D. Spiral view. □E. Peripheral view. □F. Umbilical view. □G. *Globorotalites spineus* (Cushman); umbilical view; sample 527-32-5, 33–34 cm (Maastrichtian).

chambers visible on dorsal side; 6–8 in number, gradually increasing in size as added. Dorsal sutures distinct, oblique, and curved. Ventral side (umbilical side) strongly convex, bell-shaped, and involute; umbilicus closed. Only chambers of final whorl visible on ventral side. Ventral sutures distinct, curved, and slightly depressed.

Occurrence. – Maastrichtian at Site 525, Danian and Maastrichtian at Site 527.

Remarks. – This species is similar to *Globorotalites* sp. B, but it differs in having a much larger test and less depressed ventral sutures and in being less convex and bell-shaped ventrally. Todd (1970) identified the same form under the name of *Globorotalites multiseptus* Brotzen. However, six specimens identified by Brotzen as *G. multiseptus* (USNM No. 48) were available at the Smithsonian Institution. Brotzen's specimens were distinctly different from Todd's (1970) specimens and the specimens found in the present material in having a more cone-shaped (than bell-shaped) umbilical side in side view, a distinct, open umbilicus, a much broader and

acute apertural face, and a more acute periphery. Furthermore, Brotzen's specimens are indicated by the label on the slide to be much older (lower Senonian, Eriksdal I, Sweden) than Todd's (1970) specimens and those in the present material, which all are from upper Maastrichtian strata.

Globorotalites spineus (Cushman, 1926)

Fig. 31G

Synonymy. – □v*1926a *Truncatulina spinea* n.sp. – Cushman, p. 22, Pl. 2:10. □1946 *Eponides? spinea* (Cushman) – Cushman, p. 142, Pl. 57:16. □1977 *Globorotalites spineus* (Cushman) – Sliter, p. 675, Pl. 11:8. □1978 *Globorotalites spineus* (Cushman) – Beckmann, p. 766, Pl. 4:21,27.

Material. – One single specimen.

Description. – Test trochospiral, planoconvex, bell-shaped, and irregularly subcircular in outline. Periphery very acute with a large, irregular hyaline keel, which is developed into several spines (one per chamber). Aper-

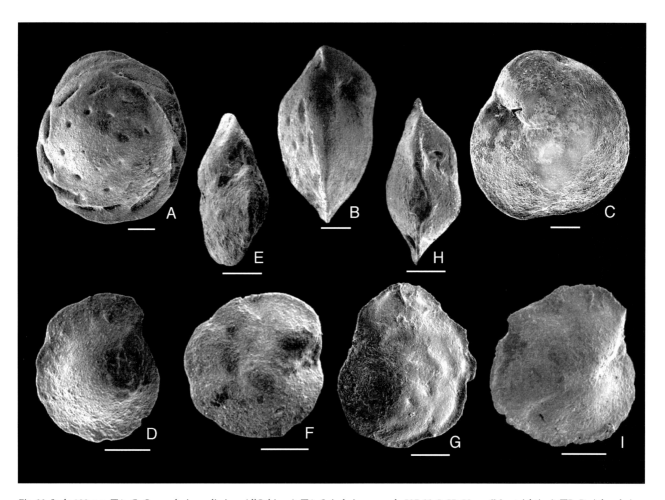

Fig. 32. Scale 100 μm. □A–C. *Osangularia cordieriana* (d'Orbigny). □A. Spiral view; sample 527-32-5, 57–58 cm (Maastrichtian). □B. Peripheral view; sample 527-32-5, 92–94 cm (Maastrichtian). □C. Umbilical view; sample 527-32-6, 30–31 cm (Maastrichtian). □D–F. *Osangularia* sp. lenticular form. □D. Spiral view; sample 527-33-1, 18–20 cm (Maastrichtian). □E. Oblique peripheral view; sample 527-32-5, 57–59 cm (Maastrichtian). □F. Umbilical view; sample 527-32-4, 7–8 cm (Maastrichtian). □G–I. *Osangularia texana* (Cushman). □G. Spiral view; sample 527-32-3, 129–130 cm (Danian). □H. Peripheral view; sample 527-32-4, 0–1 cm (Danian). □I. Umbilical view; sample 527-32-3, 129–130 cm (Danian).

ture poorly preserved, but seems to consist of an interiomarginal slit, extending from periphery to umbilicus. Walls calcareous, perforate, and smooth. Dorsal side (spiral side) flat and evolute. Chambers hardly visible on dorsal side, owing to hyaline thickening; about six chambers in final whorl. Dorsal sutures indistinct, oblique, and slightly curved. Ventral side (umbilical side) strongly convex, 'bell-shaped', and involute. Only chambers of final whorl visible on ventral side. Ventral sutures distinct, nearly straight, and slightly depressed.

Occurrence. – Maastrichtian at Site 527.

Remarks. – This very rare species is easy to recognize because of its typical, spined keel at the margin. The holotype of this species was examined at the Smithsonian Institution. The dorsal sutures was found to be obscured in the holotype, but it has short spines at the periphery, which are also present in the specimen found here.

Family Osangulariidae Loeblich & Tappan, 1964

Diagnosis. – See Loeblich & Tappan (1988, p. 629).

Genus *Osangularia* Brotzen, 1940

Type species and diagnosis. – See Loeblich & Tappan (1988, p. 630).

Osangularia cordieriana (d'Orbigny, 1840)

Fig. 32A–C

Synonymy. – □*1840 *Rotalina cordieriana* nov. sp. – d'Orbigny, p. 33, Pl. 3:9–11. □1978 *Osangularia* cf. *lens*

Brotzen – Beckmann, p. 768, Pl. 4:18–20. □1978 *Osangularia cordieriana* (d'Orbigny) – Beckmann, p. 768, Pl. 4:12–13. □1983 *Osangularia cordieriana* (d'Orbigny) – Dailey, p. 767, Pl. 6:4,7–8.

Material. – Forty specimens.

Description. – Test trochospiral, equally to unequally biconvex, and nearly circular in outline. Periphery subacute with narrow hyaline border. Walls calcareous; coarsely perforated dorsally and perforated and smooth ventrally. Aperture consisting of a small slit-like opening at base of final chamber extending up over apertural face, located halfway between periphery and umbonal boss. Dorsal side (spiral side) convex and evolute. All chambers visible on dorsal side but somewhat obscured by ornamentation; about ten chambers in final whorl in adult specimens; chambers gradually increasing in size as added. Dorsal sutures distinct, oblique, elevated, and almost straight; between elevated sutures depressed areas, which form slit-like pores in earlier formed chambers. Ventral side (umbilical side) convex and involute; large hyaline boss instead of umbilicus. Only chambers of final whorl visible on ventral side. Ventral sutures distinct, rather broad, curved, and depressed.

Occurrence. – Danian and Maastrichtian at Site 527.

Remarks. – The status of different species of *Osangularia* described in the literature on South Atlantic deep-sea benthic foraminifera is somewhat confusing. The specimen of *O. cordieriana* figured by Dailey (1983), and the specimens of *O. cordieriana*, *O.* cf. *lens*, and *O. incisa* (lenticular specimen) figured by Beckmann (1978), are difficult to separate, and they are all very close to *O. cordieriana* as distinguished here. *Osangularia cordieriana* is distinguished from other *Osangularia* encountered in the material by its keel-less periphery, nearly equal biconvexity, and strongly elevated sutures on the spiral side.

Osangularia sp. lenticular form
Fig. 32D–F

Material. – Twenty-two specimens.

Description. – Test trochospiral, unequally biconvex, compressed dorso–ventrally, and characteristically lenticular in transection; lobated and subcircular in outline. Periphery acute, but without hyaline keel. Walls calcareous, perforate, and smooth. Aperture consisting of a small slit, extending up over apertural face. Dorsal side (spiral side) convex and evolute. All chambers easily visible on dorsal side; about 7–8 chambers in the final whorl; quite rapidly increasing in size as added. Dorsal sutures distinct, narrow, straight, and nearly flush. Ventral side (umbilical side) flat to weakly convex and involute; umbilicus closed by a large, flush, hyaline boss. Only chambers of last whorl visible on ventral side. Ventral sutures distinct, curved, and depressed.

Occurrence. – Maastrichtian at Site 525, Danian and Maastrichtian at Site 527.

Remarks. – This is a very distinct form, which is separated from other *Osangularia* by its lenticular transection and keel-less periphery. The lack of the dorsal sutural hyaline thickenings, which are common in other species of *Osangularia*, is also characteristic for this species.

Osangularia texana (Cushman, 1938)
Fig. 32G–I

Synonymy. – □*v**1938 *Pulvinulinella texana* n.sp. – Cushman, p. 49, Pl. 8:8. □*v*1944a *Pulvinulinella texana* Cushman – Cushman, p. 14, Pl. 3:5. □*v*1992a *Osangularia texana* (Cushman) – Widmark & Malmgren, p. 112, Pl. 8:4.

Material. – 112 specimens.

Description. – Test trochospiral, biconvex, lenticular in transection, and subcircular in outline. Periphery acute, with narrow, hyaline keel. Walls calcareous, perforate, and relatively smooth. Aperture consisting of a small, bent slit, extending up over apertural face. Dorsal side (spiral side) weakly convex and evolute. All chambers visible on dorsal side; about seven chambers in the final whorl; chambers gradually increasing in size as added. Dorsal sutures distinct, narrow, oblique, straight to weakly curved, and flush. Ventral side (umbilical side) convex and involute; umbilicus closed by hyaline material. Only chambers of final whorl visible on ventral side. Ventral sutures distinct, slightly curved, and slightly depressed.

Occurrence. – Danian and Maastrichtian at Sites 525 and 527.

Remarks. – This species is separated from other *Osangularia* on the basis of its broad (nearly equal-sided) chambers dorsally and relatively low number of chambers (7 instead of about 10–12) in the last formed whorl. In addition to the holotype and plesiotype, some secondary types of *P. texana* Cushman (Cush. Coll. No. 40252) were examined, and they are consistent with the concept of the species applied here.

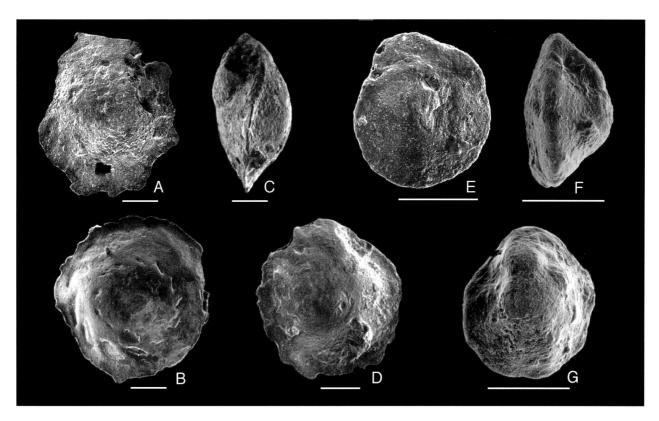

Fig. 33. Scale 100 μm. □A–D. *Osangularia velascoensis* (Cushman); sample 525A-40-1, 149–150 cm (Danian). □A. Spiral view of poorly preserved specimen. □B. Spiral view of well-preserved specimen. □C. Peripheral view. □D. Umbilical view. □E–G. *Osangularia*? spp. juvenile forms; sample 527-32-4, 19–20 cm (Danian). □E. Spiral view. □F. Peripheral view. □G. Umbilical view.

Osangularia velascoensis (Cushman, 1925)

Fig. 33A–D

Synonymy. – □*v**1925 *Truncatulina velascoensis* Cushman, n.sp. Cushman, p. 20, Pl. 3:2. □1932 *Pulvinulinella velascoensis* (Cushman) – Cushman & Jarvis, p. 48, Pl. 14:6. □*v*1992a *Osangularia velascoensis* (Cushman) – Widmark & Malmgren, p. 112, Pl. 9:1.

Material. – 100 specimens.

Description. – Test trochospiral, nearly equally biconvex, and subcircular in outline. Periphery acute; in adult specimens with thin, irregular, hyaline keel, which is often partly broken away. Walls calcareous and perforate; dorsally with large pores in the central part and depressions between elevated sutures in the last whorls, ventrally smooth. Aperture consisting of a small, bent slit, extending up over apertural face. Dorsal side (spiral side) convex and evolute. All chambers visible on dorsal side; about 10–12 chambers in final whorl, gradually increasing in size as added; chambers in juvenile stages truncated ('cog-wheel'-like). Dorsal sutures distinct in juvenile stages, in later stages the sutures become somewhat obscured; sutures oblique, broad, and flush in juvenile stages, and elevated in the adult. Ventral side (umbilical side) convex and involute; umbilicus closed by large hyaline boss, which becomes proportionally smaller as whorls are added. Only chambers of last whorl visible on ventral side. Ventral sutures distinct, relatively broad, curved, and depressed.

Occurrence. – Danian and Maastrichtian at Sites 525 and 527.

Remarks. – Most specimens here referred to this species are in a poor state of preservation. It is, however, distinguished from other *Osangularia* by its numerous chambers in the final whorl, 'cog-wheel'-like spiral side in the juvenile stages, and its thin and broad (when preserved) hyaline keel in the adult stages. Specimens of this taxon were compared with the holotype of *Truncatulina velascoensis* Cushman, and they were found to be consistent with the holotype.

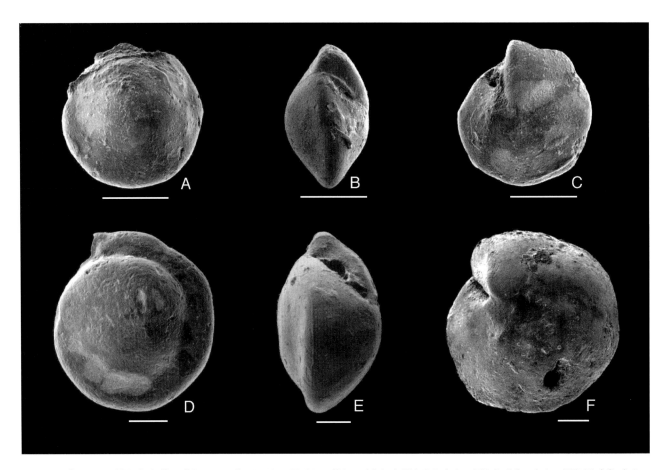

Fig. 34. Scale 100 μm. □A–C. *Oridorsalis?* sp.; sample 527-32-5, 92–94 cm (Maastrichtian). □A. Spiral view. □B. Peripheral view. □C. Umbilical view. □D–F. *Oridorsalis?* complex. □D. Spiral view; sample 525A-40-2, 78–79 cm (Maastrichtian). □E. Peripheral view; sample 525A-40-2, 59–60 cm (Maastrichtian). □F. Umbilical view; sample 525A-40-3, 130–131 cm (Maastrichtian).

Osangularia? spp. juvenile forms

Fig. 33E–G

Material. – 104 specimens.

Occurrence. – Danian and Maastrichtian at Sites 525 and 527.

Remarks. – The generic status of this taxon is uncertain. It includes small specimens with planoconvex tests and a well-developed hyaline boss.

Family Oridorsalidae Loeblich & Tappan, 1984

Diagnosis. – See Loeblich & Tappan (1988, p. 630).

Genus *Oridorsalis* Andersen, 1961

Type species and diagnosis. – See Loeblich & Tappan (1988, p. 630).

Oridorsalis? sp.

Fig. 34A–C

Material. – Sixty-two specimens.

Description. – Test trochospiral, biconvex, and circular in outline. Periphery subangular to angular. Walls calcareous, perforate, and smooth. Aperture consisting of a short interiomarginal slit, extending from periphery halfway to umbilical area. Dorsal side (spiral side) convex (more strongly in adult specimens) and evolute. All chambers visible on dorsal side; chambers in final whorl distinct, those of previous whorls obscured; about seven chambers in final whorl, slowly increasing in size as added; chambers longer than broad (2:1) in the adult specimens. Dorsal sutures distinct in final whorl, narrow, nearly orthogonal to spiral suture, straight, and slightly depressed. Ventral side (umbilical side) weakly convex and involute; umbilicus closed. Only chambers of final whorl visible on ventral side. Ventral sutures distinct, broad, nearly straight to weakly sinuate, and flush.

Occurrence. – Danian at Site 525, Danian and Maastrichtian at Site 527.

Remarks. – This taxon is fairly stable throughout the material; the chambers in spiral view tend to become shorter in the smaller, juvenile specimens. The taxonomic status of this taxon is somewhat uncertain. Its biconvex test shape and apertural features may suggest that it should be referred to *Oridorsalis*, but the sutural openings near the junction of septal and spiral sutures, typical for this genus, could not be observed with certainty throughout the specimens analyzed. Specimens referred to *Oridorsalis*? sp. are separated from the specimens included in the *Oridorsalis*? complex on the basis of their generally smaller test size, straight dorsal sutures, longer chambers in spiral view, and more transparent, perforate test walls.

Oridorsalis? complex

Fig. 34D–F

Material. – 158 specimens.

Occurrence. – Danian and Maastrichtian at Sites 525 and 527.

Remarks. – This taxon includes several *Oridorsalis*-like morphotypes, which are difficult to separate consistently, because of their overlapping morphologies. Specimens of the *Oridorsalis*? complex grade from equally to unequally biconvex test shapes; they have shiny, opaque (finely perforated to pore-less) test walls, curved to oblique dorsal sutures, curved to sigmoid ventral sutures, and an interiomarginal aperture.

Family Gavelinellidae Hofker, 1956

Diagnosis. – See Loeblich & Tappan (1988, p. 633).

Subfamily Gyroidinoidinae Saidova, 1981

Diagnosis. – See Loeblich & Tappan (1988, p. 633).

Genus *Gyroidinoides* Brotzen, 1942

Type species and diagnosis. – See Loeblich & Tappan (1988, p. 633).

Gyroidinoides beisseli (White, 1928)

Fig. 35A–C

Synonymy. – □*1928b *Gyroidina beisseli* n.sp. – White, p. 291, Pl. 39:7. □1946 *Gyroidina beisseli* White – Cushman, p. 141, Pl. 58:11. □1977 *Gyroidinoides beisseli* (White) – Sliter, p. 675, Pl. 10:3–6. □1978 *Gyroidina beisseli* White – Beckmann, p. 767, Pl. 3:29–30. □1983 *Gyroidinoides beisseli* (White) – Dailey, p. 767, Pl. 6:5–6,9. □v1992a *Gyroidinoides beisseli* (White) – Widmark & Malmgren, p. 7, Pl. 7:2. □v1992b *Gyroidinoides beisseli* (White) – Widmark & Malmgren, p. 394, Pl. 3:7–9.

Material. – 190 specimens.

Description. – Test trochospiral and nearly planoconvex; weakly lobated and subcircular in outline. Periphery rounded to subangular. Walls calcareous, finely perforated, and smooth. Aperture consisting of interiomarginal slit, located at middle of base of last chamber. Dorsal side (spiral side) weakly convex and evolute. All chambers visible on dorsal side; slightly inflated and gradually increasing in size as added. Dorsal sutures distinct, quite broad, oblique, nearly straight, and flush to very slightly depressed. Ventral side (umbilical side) convex and involute; umbilicus almost closed. Only chambers of final whorl visible on ventral side. Ventral sutures distinct, broad around umbilicus, sinuate, and flush to slightly depressed.

Occurrence. – Danian and Maastrichtian at Site 527.

Remarks. – Specimens referred to this species have a pronouncedly convex umbilical side and a weakly convex spiral side; they are similar to the specimens here identified as *G. goudkoffi* (Trujillo), but they differ in having a less acute periphery and a less developed umbilicus. Some secondary types (Cush. Coll. No. 46769) referred to this species were available at the Smithsonian Institution and examined, and they were found to be conspecific with the specimens here identified with this species.

Gyroidinoides globosus (Hagenow, 1842)

Fig. 35H–J

Synonymy. – □*1842 *Nonionina globosa* von Hagenow n.sp. – Hagenow, p. 574. □v1931c *Gyroidina globosa* (Hagenow) – Cushman, p. 310, Pl. 35:19. □v1932 *Gyroidina globosa* (Hagenow) – Cushman & Jarvis, p. 47, Pl. 14:4–5. □v1944a *Gyroidina globosa* (Hagenow) –

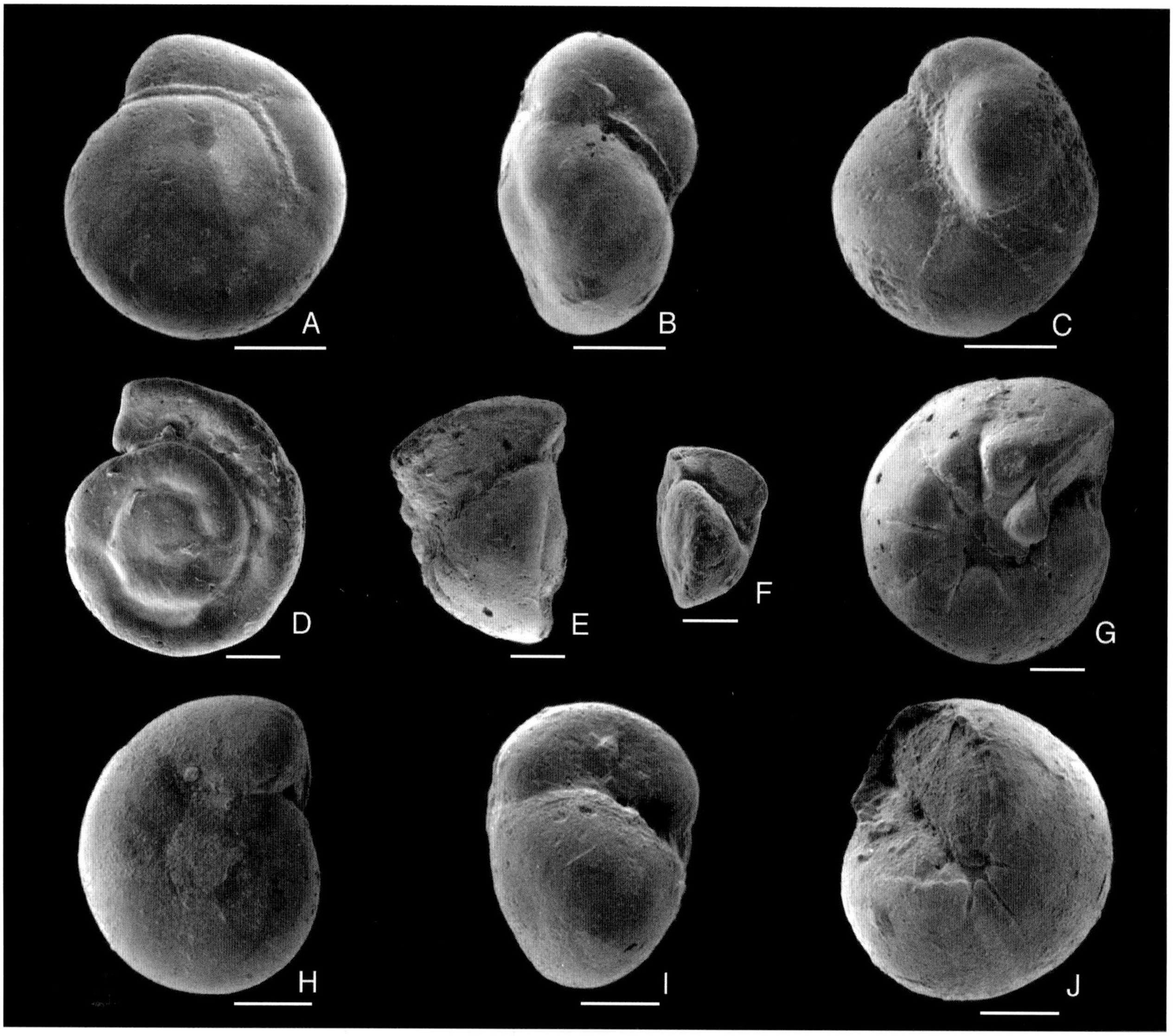

Fig. 35. Scale 100 μm. □A–C. *Gyroidinoides beisseli* (White); sample 527-32-5, 92–94 cm (Maastrichtian). □A. Spiral view. □B. Peripheral view. □C. Umbilical view. □D–G. *Gyroidinoides tellburmaensis* Futyan. □D. Spiral view (adult specimen); sample 525A-40-2, 109–110 cm (Maastrichtian). □E. Peripheral view (adult specimen); sample 525A-40-2, 92–93 cm (Maastrichtian). □F. Peripheral view (juvenile specimen); sample 525A-40-2, 78–79 cm (Maastrichtian). □G. Umbilical view (adult specimen); sample 525A-40-2, 78–79 cm (Maastrichtian). □H–J. *Gyroidinoides globosus* (Hagenow). □H. Spiral view; sample 527-32-4, 87–89 cm (Maastrichtian). □I. Peripheral view; sample 525A-40-2, 47–48 cm (Maastrichtian). □J. Umbilical view; sample 527-32-4, 87–89 cm (Maastrichtian).

Cushman, p. 13, Pl. 3:3. □*v*1946 *Gyroidina globosa* (Hagenow) – Cushman & Renz, p. 44, Pl. 7:15. □*v*1956 *Gyroidina globosa* Hagenow – Said & Kenawy, p. 149, Pl. 5:5. □1983 *Gyroidinoides globosus* (Hagenow) – Dailey, p. 767, Pl. 7:4,7–8. □1986 *Gyroidinoides globosus* (Hagenow) – Van Morkhoven *et al.*, p. 326, Pl. 107. □1991 *Gyroidinoides globosus* (Hagenow) – Nomura, p. 22, Pl. 4. □1992 *Gyroidinoides globosus* (Hagenow) – Gawor-Biedowa, p. 155, Pl. 32:7–9.

Material. – Fifteen specimens.

Description. – Test trochospiral, subglobular, and subcircular in outline. Periphery broadly rounded. Walls calcareous, finely perforate, and smooth. Aperture consisting of an interiomarginal slit, extending from periphery to umbilicus; large apertural face. Dorsal side (spiral side) convex and evolute (sometimes the last whorl overlaps the previous one dorsally). All chambers visible on dorsal side, but somewhat obscured in the inner whorls; about six chambers in final whorl. Dorsal sutures distinct (at least in the final whorl), quite broad, nearly straight and orthogonal to spiral suture, and depressed. Ventral side (umbilical side) strongly convex and involute; umbilicus small and open. Only chambers of final whorl visible on ventral side. Ventral sutures distinct, quite broad, nearly straight, and very slightly depressed.

Occurrence. – Danian and Maastrichtian at Site 525, Danian at Site 527.

Remarks. – This species occurs in very low frequencies in a few samples only. *Gyroidinoides globosus* is recognized by its subglobular test shape and relatively large test size. Several secondary types of *Gyroidina globosa* (given above) were available at the Smithsonian Institution and examined, and they are regarded as consistent with the concept of *G. globosus* (Hagenow) applied here. However, some topotypes of this species from the Upper Cretaceous (Santonian, Campanian) chalk of the Island of Rügen were also examined at the Smithsonian Institution. The topotypes were found to be much larger than the specimens in the present material, but they were similar in having about six chambers in the final whorl, straight sutures both dorsally and ventrally, and a small, open umbilicus.

Gyroidinoides goudkoffi (Trujillo, 1960)

Fig. 36A–C

Synonymy. – □*1960 *Eponides goudkoffi* n.sp. – Trujillo, p. 333, Pl. 48:6. □1983 *Gyroidinoides goudkoffi* (Trujillo) – Dailey, p. 767, Pl. 7:5–6,9. □1991 *Oridorsalis* sp. 1 – Nomura, Pl. 5:6.

Material. – Sixty specimens.

Description. – Test trochospiral, nearly planoconvex, and subcircular in outline. Periphery angular. Walls calcareous, finely perforate, and smooth. Aperture consisting of an interiomarginal slit, extending from halfway of apertural face to umbilicus; apertural face large. Dorsal side (spiral side) nearly flat, with central area vaulted; evolute. All chambers visible on dorsal side, but somewhat obscured in the inner whorls; about six chambers in final whorl. Dorsal sutures indistinct, curved, and depressed. Ventral side (umbilical side) strongly convex and involute; umbilicus small and open. Only chambers of final whorl visible on ventral side. Ventral sutures distinct, broad, sinuate, and flush.

Occurrence. – Danian and Maastrichtian at Sites 525 and 527.

Remarks. – This species is similar to *G. beisseli*, but it differs in having an acute periphery and a small, open umbilicus.

Gyroidinoides nitidus (Reuss, 1845)

Fig. 36D–F

Synonymy. – □*1845 *Rotalina nitida* nov. sp. – Reuss, p. 35, Pl. 8:52; 12:8, 20. □1928b *Gyroidina nitida* (Reuss) – White, p. 296, Pl. 40:6. □1934 *Gyroidina nitida* (Reuss) – Morrow, p. 197, Pl. 30:1. □1983 *Gyroidinoides nitidus* (Reuss) – Dailey, p. 767, Pl. 7:10–12.

Material. – Fifty-six specimens.

Description. – Test small, trochospiral, unequally biconvex, and subcircular in outline. Periphery broadly rounded. Walls calcareous, perforate, and smooth. Aperture consisting of an interiomarginal slit, extending from periphery to umbilicus; large apertural face. Dorsal side (spiral side) weakly convex and evolute. All chambers visible on dorsal side; about six slightly inflated chambers in final whorl; chambers increasing gradually in size as added. Dorsal sutures distinct, narrow, depressed, and nearly orthogonal to spiral. Ventral side (umbilical side) strongly convex and evolute; umbilicus distinct and open. Only chambers of final whorl visible on ventral side. Ventral sutures distinct, narrow, straight to weakly sinuous, and slightly depressed.

Occurrence. – Danian and Maastrichtian at Site 525, Maastrichtian at Site 527.

Remarks. – This species is similar to *G. globosa*, but it may be differentiated on the basis of its much smaller test size, more compressed test, and more perforated test walls. Several secondary types, named *Gyroidina nitida* Reuss, at the Smithsonian Institution were examined. A topotype from the Upper Cretaceous (Turonian) 'Plänarmergel' (Dresden, Germany) was examined, and it was found to be poorly preserved, but in agreement with the concept of *G. nitidus* (Reuss) used in this communication with regard to test size, chamber shape, the open umbilicus, and the nature of the sutures. However, examination of a number of other secondary types filed under this species at the Cushman Collection revealed that the concept of the species has varied considerably. These secondary types ranged morphologically between forms similar to *G. tellburmaensis* (Futyan) (with acute periphery and deeply convex umbilical side) and equally biconvex forms with closed umbilici and strongly sinuous ventral sutures, approaching the morphological features of *Eponides* or *Oridorsalis*. A large number of *Gyroidina nitida* (Reuss) was also available at Brotzen's collection at the Naturhistoriska Rikmuseet, Stockholm. Examination of Brotzen's specimens led to the same conclusion as the examination of the types at the Cushman Collection. It is obvious that the original concept of *Rotalia nitida* Reuss [= *Gyroidinoides nitidus* (Reuss)] was widely defined and that it needs some revision if it is to be possible to use it consistently in the future.

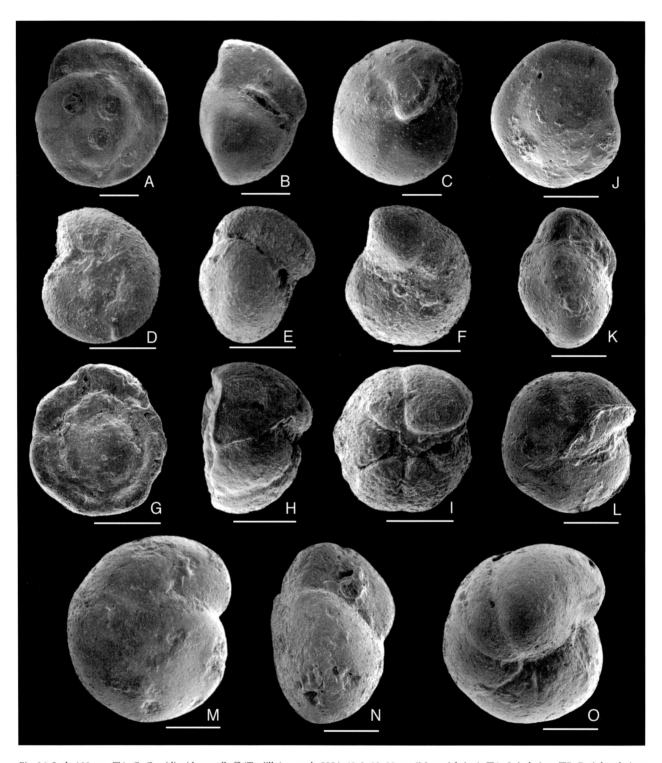

Fig. 36. Scale 100 µm. □A–C. *Gyroidinoides goudkoffi* (Trujillo); sample 525A-40-2, 92–93 cm (Maastrichtian). □A. Spiral view. □B. Peripheral view. □C. Umbilical view. □D–F. *Gyroidinoides nitidus* (Reuss); sample 525A-40-2, 92–93 cm (Maastrichtian). □D. Spiral view. □E. Peripheral view. □F. Umbilical view. □G–I. *Gyroidinoides quadratus* (Cushman & Church); sample 525A-40-3, 70–71 cm (Maastrichtian). □G. Spiral view. □H. Peripheral view. □I. Umbilical view. □J–L. *Gyroidinoides* sp. C; sample 527-32-4, 19–20 cm (Danian). □J. Spiral view. □K. Peripheral view. □L. Umbilical view. □M–O. *Gyroidinoides* spp. □M. Spiral view; sample 527-32-3, 129–130 cm (Danian). □N. Peripheral view; sample 527-32-4, 0–1 cm (Danian). □O. Umbilical view; sample 527-32-4, 0–1 cm (Danian).

Gyroidinoides quadratus (Cushman & Church, 1929)

Fig. 36G–I

Synonymy. – □*1929 *Gyroidina quadrata* n.sp. – Cushman & Church, p. 516, Pl. 41:7–9. □1964 *Gyroidina quadrata* Cushman & Church – McGugan, p. 944, Pl. 151:5. □1977 *Gyroidinoides quadratus* (Cushman & Church) – Sliter, p. 675, Pl. 11:4–5,7. □1978 *Gyroidina quadrata* Cushman & Church – Beckmann, p. 767, Pl. 4:6. □1983 *Gyroidinoides quadratus* (Cushman & Church) – Dailey, p. 767, 8:2–4. □1983 *Gyroidinoides quadratus* (Cushman & Church) – Tjalsma & Lohmann, p. 15, Pl. 5:6. □v1988 *Gyroidinoides quadratus* (Cushman & Church) – Widmark & Malmgren, p. 73, Pl. 4:2. □1991 *Gyroidinoides quadratus* (Cushman & Church) – Nomura, p. 22, Pl. 4:7. □1992 *Gyroidinoides quadratus* (Cushman & Church) – Kaiho, p. 255, Pl. 4:6. □v1992a *Gyroidinoides quadratus* (Cushman & Church) – Widmark & Malmgren, p. 111, Pl. 6:4.

Material. – 103 specimens.

Description. – Test trochospiral, concavoconvex, subcircular in outline. Periphery acute, elevated above central spire dorsally. Walls calcareous, smooth, and perforate. Aperture consisting of an interiomarginal slit at base of last chamber, extending from periphery to umbilicus. Outer margin of dorsal side (spiral side) elevated above central spire, which is slightly vaulted; circular and lobated in outline; evolute. All chambers visible on dorsal side; chambers long and narrow. Dorsal sutures distinct, depressed, and oblique in the final whorl; sutures obscured in the inner whorls. Ventral side (umbilical side) strongly convex, circular, and lobated in outline; small umbilicus, which may be partly infilled with calcitic material; involute. Only chambers of the final whorl visible on ventral side; about six tulip-petal-shaped chambers in the last whorl. Ventral sutures distinct, narrow, straight, and depressed.

Occurrence. – Danian at Site 525, Danian and Maastrichtian at Site 527.

Remarks. – This species is easily recognizable on the basis of its concavoconvex outline in side view and its typical tulip-petal-shaped chambers in the last whorl of the umbilical side. Unfortunately, no types filed under this species were available at the Smithsonian Institution.

Gyroidinoides sp. C

Fig. 36J–L

Material. – 120 specimens.

Description. – Test trochospiral, unequally biconvex and subcircular in outline. Periphery rounded. Walls calcareous, perforate, and smooth. Aperture consisting of an interiomarginal slit, extending from periphery to umbilicus. Dorsal side (spiral side) weakly convex and evolute. All chambers visible on dorsal side; chambers of final whorl about twice as long as broad and about five in number; chambers increasing gradually in size as added. Dorsal sutures distinct, narrow, slightly depressed, nearly straight, and almost orthogonal to spiral. Ventral side (umbilical side) strongly convex and involute; umbilicus closed. Only chambers of final whorl visible on ventral side. Ventral sutures distinct, rather broad, strongly sinuate, and flush.

Occurrence. – Danian and Maastrichtian at Site 527.

Remarks. – This species resembles *G. beisseli*, but it differs in having orthogonal sutures on the spiral side instead of oblique spiral sutures as in *G. beisseli*. In side view, the periphery is more rounded in *Gyroidinoides* sp. C than in *G. beisseli*, and the chambers of the last whorl in dorsal view are longer in *Gyroidinoides* sp. C than in *G. beisseli* (the ratio length/breadth is 2:1 and nearly 1:1, respectively). *Gyroidinoides* sp. C is similar to the specimen of *Gyroidina naranjoensis* n.sp. figured by White (1928b), but it differs from White's species in having more strongly sinuate ventral sutures.

Gyroidinoides spp.

Fig. 36M–O

Material. – 289 specimens.

Occurrence. – Danian and Maastrichtian at Sites 525 and 527.

Remarks. – Several morphotypes of *Gyroidinoides*, which are difficult to separate, are included in this taxon. They have a nearly flat spiral side, a strongly convex umbilical side, limbate sutures, and rapidly increasing chambers in the last whorl. Some specimens of *Gyroidinoides* spp. show some resemblance to the specimen of *G. octocamerata* (Cushman & Hanna) identified by Beckmann (1978).

Gyroidinoides tellburmaensis Futyan, 1976

Fig. 35D–G

Synonymy. – □v1944b *Gyroidina girardana* (Reuss) – Cushman, p. 95, Pl. 14:24. □v1946 *Gyroidina girardana* (Reuss) – Cushman & Renz, p. 44, Pl. 7:20. □v1956 *Gyroidina girardana* (Reuss) – Said & Kenawy, p. 148, Pl. 5:7. □*1976 *Gyroidinoides tellburmaensis* n.sp. – Futyan, p. 532, Pl. 81:10–12. □1983 *Gyroidinoides girardanus*

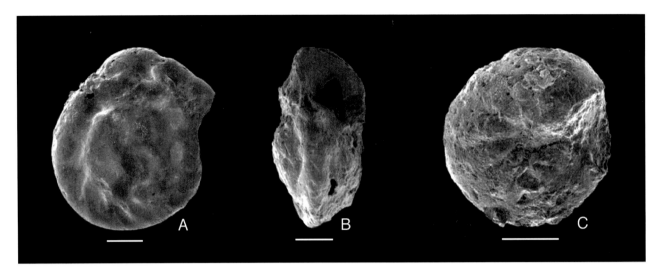

Fig. 37. Scale 100 μm. □A–C. *Angulogavelinella avnimelechi* (Reiss). □A. Spiral view; sample 525A-40-1, 130–131 cm (Danian). □B. Peripheral view; sample 525A-40-1, 140–141 cm (Danian). □C. Umbilical view; sample 525A-40-1, 140–141 cm (Danian).

(Reuss) – Dailey, p. 767, Pl. 7:1–3. □1992 *Gyroidinoides girardanus* (Reuss) – Gawor-Biedowa, p. 154, Pl. 32:10–12. □v1992a *Gyroidinoides girardanus* (Reuss) – Widmark & Malmgren, p. 111, Pl. 6:3. □v1994 *Gyroidinoides tellburmaensis* Futyan – Speijer, p. 62, Pl. 3:1.

Material. – Seventy-seven specimens.

Description. – Test trochospiral, planoconvex, and subcircular in outline. Periphery subangular. Walls calcareous, finely perforate, and smooth. Aperture consisting of an interiomarginal slit, extending from periphery to umbilicus; large, high, and flat apertural face. Dorsal side (spiral side) flat and evolute. All chambers visible on dorsal side; about eight chambers in final whorl; chambers increasing slowly in size as added. Dorsal sutures not very distinct, especially those of inner whorls; sutures oblique, slightly curved, and flush or very slightly depressed. Ventral side (umbilical side) convex and involute; umbilicus open and deep. Only chambers of final whorl visible on ventral side. Ventral sutures distinct, nearly straight, and depressed (flush in juvenile specimens).

Occurrence. – Danian and Maastrichtian at Site 525, Maastrichtian at Site 527.

Remarks. – This species is distinguished from other *Gyroidinoides* on the basis of its subangular periphery, distinct ventral sutures, typical apertural face, and distinct umbilicus. The specimens here referred to this species were found to be consistent with the examined secondary types of *Gyroidina girardana* (Reuss) given in the synonymy list and has usually been referred to that Reuss species throughout the literature. The Reuss and Futyan species are different, however, in that the former has orthogonal (instead of oblique) sutures dorsally and a much more rounded (instead of a subacute) periphery.

Subfamily Gavelinellinae Hofker, 1956

Diagnosis. – See Loeblich & Tappan (1988, p. 635).

Genus *Angulogavelinella* Hofker, 1957

Type species and diagnosis. – See Loeblich & Tappan (1988, p. 635).

Angulogavelinella avnimelechi (Reiss, 1952)

Fig. 37A–C

Synonymy. – □*1952 *Pseudovalvulineria avnimelechi* n.sp. – Reiss, p. 269, Pl. 2. □1986 *Angulogavelinella avnimelechi* (Reiss) – Van Morkhoven *et al.*, p. 344, Pl. 112. □v1992a *Stensioeina* sp. A (Reiss) – Widmark & Malmgren, p. 113, Pl. 9:3. □v1994 *Angulogavelinella avnimelechi* (Reiss) – Speijer, p. 166, Pl. 3:7.

Material. – Fifty-seven specimens.

Description. – Test trochospiral, nearly planoconvex, and subcircular in outline. Periphery angular. Walls calcareous and coarsely perforated on ventral side, small ridges of secondary calcite between the sutures radiate from

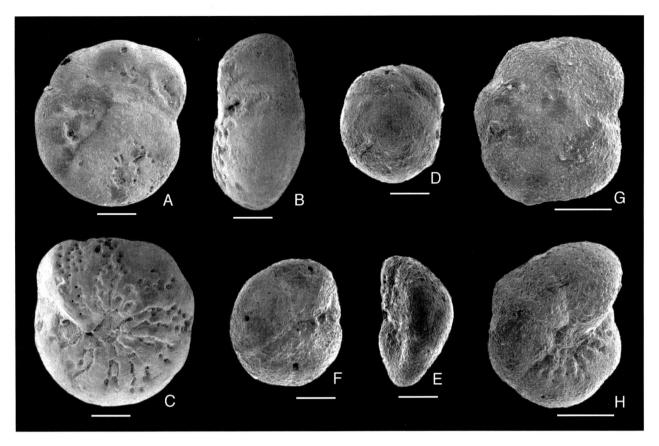

Fig. 38. Scale 100 µm. □A–C. *Gavelinella beccariiformis* (White); sample 525A-40-2, 47–48 cm (Maastrichtian). □A. Spiral view. □B. Peripheral view. □C. Umbilical view. □D–F. *Gavelinella?* sp. conical form. □D. Spiral view; sample 527-32-3, 129–130 cm (Danian). □E. Peripheral view; sample 527-32-3, 129–130 cm (Danian). □F. Umbilical view; sample 527-32-4, 0–1 cm (Danian). □G–H. *Gavelinella* spp. lobated forms; sample 527-32-4, 57–59 cm (Maastrichtian). □G. Spiral view. □H. Umbilical view.

umbilical area toward periphery; dorsal side smooth and almost poreless. Aperture consisting of an interiomarginal slit, extending from periphery to umbilicus. Dorsal side (spiral side) flat to weakly convex; evolute. All chambers visible on dorsal side; about ten chambers in final whorl. Dorsal sutures distinct, slightly curved, and elevated. Ventral side (umbilical side) convex and involute; umbilicus open. Only chambers of final whorl visible on ventral side. Ventral sutures distinct, depressed, curved, and limbate.

Occurrence. – Danian and Maastrichtian at Site 525, Maastrichtian at Site 527.

Remarks. – This distinct species has a ventral side similar to that of *Gavelinella beccariiformis*. The two species differ, however, in the nature of their dorsal sutures, which are elevated in *A. avnimelechi* but flush to depressed in *G. beccariiformis*. In addition, *A. avnimelechi* has narrower and more numerous chambers and a less convex ventral side compared to *G. beccariiformis*.

Genus *Gavelinella* Brotzen, 1942

Type species and diagnosis. – See Loeblich & Tappan (1988, p. 638).

Gavelinella beccariiformis (White, 1928)

Fig. 38A–C

Synonymy. – □*1928b *Rotalia beccariiformis* var. – White, p. 287, Pl. 39:3–4. □v1940 *Gyroidina infrafossa* n.sp. – Finlay, p. 462, Pl. 66:181–183. □v1946 *Anomalina beccariiformis* (White) – Cushman & Renz, p. 48, Pl. 8:21–22. □v1947 *Anomalina beccariiformis* (White) – Cushman & Renz, p. 50, Pl. 12:14. □1964 *Anomalina whitei* nov.sp. – Martin, p. 106, Pl. 16:4. □1977 *Gavelinella whitei* (Martin) – Sliter, p. 675, Pl. 13:2–5. □1978 *Gavelinella beccariiformis* (White) – Beckmann, p. 766, Pl. 5:1–3. □1978 *Gavelinella beccariiformis* (White) – Proto Decima & Bolli, p. 793, Pl. 6:3–4. □1983 *Gavelinella beccariiformis* (White) – Dailey, p. 766, Pl. 9:4,9. □1983 *Gavelinella beccariiformis* (White) – Tjalsma & Lohmann, p. 12, Pl. 6:1–

3. □1984 *Gavelinella beccariiformis* (White) – Clark & Wright, p. 464, Pl. 6:3. □1984 *Gavelinella beccariiformis* (White) – Nyong & Olsson, p. 473, Pl. 6:11–12. □1984 *Gavelinella whitei* (Martin) – Nyong & Olsson, p. 473, Pl. 6:7–8. □1986 *Gavelinella beccariiformis* (White) – Van Morkhoven *et al.*, p. 346, Pl. 113. □*v*1988 *Gavelinella beccariiformis* (White) – Widmark & Malmgren, p. 75, Pl. 3:5. □1990 *Gavelinella beccariiformis* (White) – Thomas, p. 590, Pl. 3:5. □1991 *Stensioeina beccariiformis* (White) – Nomura, p. 23, Pl. 1:8–9. □1992 *Stensioeina beccariiformis* (White) – Gawor-Biedowa, p. 157, Pl. 37:9–11. □*v*1992a *Gavelinella beccariiformis* (White) – Widmark & Malmgren, p. 111, Pl. 5:3. □*v*1992b *Gavelinella beccariiformis* (White) – Widmark & Malmgren, p. 393, Pl. 2:7–9. □*v*1994 *Gavelinella beccariiformis* (White) – Speijer, p. 62, Pl. 10:4.

Material. – About 1500 specimens.

Description. – Test trochospiral and planoconvex to unequally biconvex; slightly lobated and subcircular in outline. Periphery rounded. Aperture consisting of an interiomarginal slit extending from periphery to umbilicus. Walls calcareous and perforate; smooth dorsally and more coarsely perforated ventrally, where small ridges between the sutures radiate from umbilical area toward periphery. Dorsal side (spiral side) flat to slightly convex and evolute. All dorsal chambers visible; increasing gradually in size as added. Dorsal sutures distinct, depressed, and curved. Ventral side (umbilical side) more convex than dorsal side and involute; umbilicus covered by a small flap (= extension of last chamber), which may be broken in some specimens. Only chambers of last whorl visible ventrally. Ventral sutures distinct, broad, curved, and depressed.

Occurrence. – Danian and Maastrichtian at Sites 525 and 527.

Remarks. – Together with *Nuttallides truempyi*, this species is the dominating constituent of the 'Velasco-type' fauna. *Gavelinella beccariiformis* is characteristic of the Maastrichtian–Paleocene deep-sea deposits from the South Atlantic, and it is the most abundant species in the material studied. In the Paleocene, *G. beccariiformis* represents a member of a 'relict' Mesozoic deep-sea fauna, which became extinct near the Paleocene–Eocene boundary (Tjalsma & Lohmann 1983). *Gavelinella beccariiformis* was not reported by Sliter (1977), but he reported a similar species, *G. whitei* (Martin). This species was established as *Anomalina whitei* Martin by Martin (1964) on the basis of one of the varieties of *Rotalia beccariiformis* White (1928; Pl. 39:4). Tjalsma & Lohmann (1983) were of the opinion that the typical thread-like depressions around the umbilicus of *G. whitei* may be due to poor preservation. Finlay (1940) described a similar species under the name of *Gyroidina infrafossa*, which was distin-

guished from *G. beccariiformis* (White) in being heavier ornamented. Examination of the holotype of Finlay's species led to the conclusion that Finlay's species is a junior synonym of *G. beccariiformis* (White); the heavier ornamentation that Finlay (1940) pointed out seems to be well within the intraspecific variation of *G. beccariiformis* (White) and the species concept used here for this species.

Gavelinella spp. lobated forms

Fig. 38G–H

Material. – Thirty-four specimens.

Occurrence. – Danian and Maastrichtian at Sites 525 and 527.

Remarks. – A relatively low number of gavelinelliid specimens with lobated outlines were referred to this taxon.

Gavelinella? sp. conical form

Fig. 38D–F

Material. – Thirty-four specimens.

Description. – Test trochospiral, planoconvex to unequally biconvex; circular in outline. Periphery subangular. Aperture consisting of an interiomarginal slit, extending from periphery to around umbonal boss. Walls calcareous; finely perforated and smooth dorsally; coarsely perforated ventrally. Dorsal side (spiral side) convex; coiling not completely evolute. All dorsal chambers visible and arranged into two whorls only; about ten chambers in final whorl. Dorsal sutures distinct, broad, nearly straight, and flush. Ventral side (umbilical side) flat to weakly convex and involute; umbilicus completely closed by hyaline plug. Only chambers of final whorl visible ventrally. Ventral sutures distinct, broad, nearly straight, and flush.

Occurrence. – Maastrichtian at Site 525, Danian and Maastrichtian at Site 527.

Remarks. – The taxonomic status of this form is uncertain. It is provisionally placed within *Gavelinella*, because of its resemblance to the specimens of *G. beccariiformis* (White), conical variety figured by Beckmann (1978).

Genus *Paralabamina* Hansen, 1970

Type species and diagnosis. – See Loeblich & Tappan (1988, p. 641).

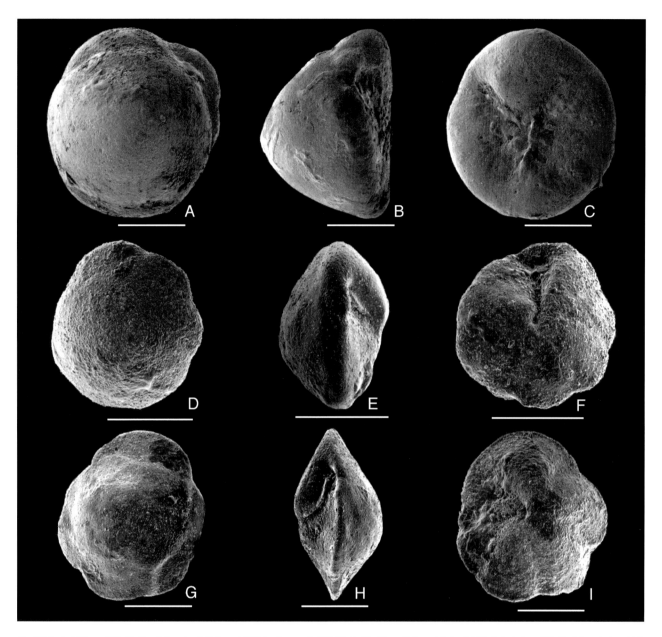

Fig. 39. Scale 100 µm. □A–C. *Paralabamina hillebrandti* (Fisher); sample 527-32-4, 87–89 cm (Maastrichtian). □A. Spiral view. □B. Peripheral view. □C. Umbilical view. □D–F. *Paralabamina* sp. intermediate form; sample 527-32-4, 58–59 cm (Maastrichtian). □D. Spiral view. □E. Peripheral view. □F. Umbilical view. □G–I. *Paralabamina lunata* (Brotzen); sample 527-32-4, 58–59 cm (Maastrichtian). □G. Spiral view. □H. Peripheral view. □I. Umbilical view.

Paralabamina hillebrandti (Fisher, 1969)

Fig. 39A–C

Synonymy. – □[?]1928b *Rotalia* cf. *partschiana* (d'Orbigny) – White, p. 288, Pl. 38:10. □[*non*] 1936 *Eponides whitei* n.sp. – Brotzen, p. 167, Pl. 12:5–8. □1962 *Eponides whitei* n.sp. – von Hillebrandt, p. 106, Pl. 8:11. □*1969 *Neoeponides hillebrandti* nom. nov. – Fisher, p. 196. □1977 *Gavelinella cayeuxi mangshlakensis* (Vasilenko) – Sliter, p. 675, Pl. 12:1–3. □1978 *Conorbina*

cf. *marginata* Brotzen – Beckmann, p. 765, Pl. 2:20–22. □1983 *Neoeponides hillebrandti* Fisher – Dailey, p. 767, Pl. 3:11–12. □1983 *Neoeponides hillebrandti* Fisher – Tjalsma & Lohmann, p. 16, Pl. 7:9. □*v*1988 *Neoeponides hillebrandti* Fisher – Widmark & Malmgren, p. 71, Pl. 2:3. □1990 *Neoeponides hillebrandti* Fisher – Thomas, p. 590, Pl. 2:3–4. □*v*1992a *Paralabamina hillebrandti* (Fisher) – Widmark & Malmgren, p. 112, Pl. 3:1. □*v*1992b *Paralabamina hillebrandti* (Fisher) – Widmark & Malmgren, p. 398, Pl. 4:7–9.

Material. – Ninety specimens.

Description. – Test trochospiral, planoconvex to unequally biconvex; circular in outline. Periphery rounded, with narrow hyaline border. Walls calcareous; smooth, finely perforated, and transparent dorsally; coarsely perforate ventrally, where umbonal area is covered with anastomosing pillars and ridges, small curved ridges radiate between sutures from umbonal cover to periphery. Aperture consisting of a simple arc at depression in apertural face, extending into umbonal area. Dorsal side (spiral side) strongly convex (nearly cone-shaped) and evolute; chambers numerous (35–40), small, all of them visible and shiny on dorsal side. Dorsal sutures distinct, flush, curved, and covered by layers of hyaline material, which gives the surface a smooth impression. Ventral side (umbilical side) flat to weakly convex and centroumbonate; involute. Only chambers of last whorl visible on ventral side. Ventral sutures flush, curved, and obscured by layers of hyaline material.

Occurrence. – Danian and Maastrichtian at Sites 525 and 527.

Remarks. – Fisher (1969) renamed this species *Neoeponides hillebrandti*, since its original name, *Eponides whitei* (von Hillebrandt, 1964) was preoccupied. The specimen figured by Dailey (1983) has a more strongly convex umbilical side than the specimens here referred to this species. The specimens found here show close resemblance to the figures of *N. hillebrandti* Fisher shown by Tjalsma & Lohmann (1983), as well as to the figured specimens of *Gavelinella cayeuxi mangshlakensis* (Vasilenko) given by Sliter (1977), and *Conorbina* cf. *marginata* Brotzen in Beckmann (1978). Two syntypes of *Conorbina marginata* Brotzen (USNM No:s P54 and P3094) were available at the Cushman Collection and examined. These syntypes were from lowest Senonian of Eriksdal, Sweden (given on slide), and identified by Brotzen, and found to be far outside the concept of *hillebrandti* used in this communication. They differ from the specimens in the material in having an acute periphery, a lobated outline, and in being distinctly cone-shaped. The generic position of this species has been regarded as uncertain, as pointed out by Tjalsma & Lohmann (1983). However, it is here referred to *Paralabamina* on the basis of its similarities to *P.* sp. intermediate form and to the type species of this genus, *P. lunata* (Brotzen).

Paralabamina lunata (Brotzen, 1948)

Fig. 39G–I

Synonymy. – □*1948 *Eponides lunata* n.sp. – Brotzen, p. 77, Pl. 10:17–18. □1970 *Eponides* sp. B – Todd, p. 145, Pl.

3:2. □1983 *Neoeponides lunata* (Brotzen) – Dailey, p. 767, Pl. 4:1–2. □v1988 *Neoeponides lunata* (Brotzen) – Widmark & Malmgren, p. 71, Pl. 2:5. □v1992a *Paralabamina lunata* (Brotzen) – Widmark & Malmgren, p. 112, Pl. 3:3. □v1992b *Paralabamina lunata* (Brotzen) – Widmark & Malmgren, p. 402, Pl. 4:10–12.

Material. – 125 specimens.

Description. – Test trochospiral, planoconvex to weakly biconvex, and lenticular in section; subcircular and lobated in outline. Periphery acute with hyaline border or narrow keel. Walls calcareous, smooth, and finely perforate. Aperture consisting of interiomarginal slit extending from marginal depression at base of last chamber to umbilicus. Dorsal side (spiral side) nearly flat with a vaulted central area and flattening toward the margin; evolute. All chambers visible on dorsal side; chambers of last whorl crescent-shaped and about three times longer than broad. Dorsal sutures distinct, broad, and consisting of hyaline material, which gives the sutures a lustrous appearance. Ventral side (umbilical side) convex and involute, with open umbilicus. Only chambers of final whorl visible on ventral side; about six in number. Ventral sutures distinct, radial, depressed, and slightly curved.

Occurrence. – Danian and Maastrichtian at both Site 525 and 527.

Remarks. – Only specimens that closely resemble the figure of the holotype of *Eponides lunata* (Brotzen, 1948) are included in this species. Brotzen's species is nearly planoconvex, with an almost flat spiral side and a strongly convex umbilical side. Several specimens of *Eponides lunata* Brotzen were available at the Naturhistoriska Riksmuseet, Stockholm, and they agree very closely with the specimens found here. Hansen (1970) established a new genus, *Paralabamina*, for Brotzen's species; this genus is also used here for Fisher's *hillebrandti* and the 'intermediate form'. *Paralabamina lunata* is previously reported from the Maastrichtian of the South Atlantic by Todd (1970), who named it *Eponides* sp. B. Dailey (1983) and Sliter (1977) reported this species under the generic names of *Neoeponides* and *Gyroidinoides*, respectively.

Paralabamina sp. intermediate form

Fig. 39D–F

Synonymy. – □1983 *Neoeponides* cf. *lunata* (Brotzen) – Tjalsma & Lohmann, p. 16, Pl. 7:10. □v1988 *Neoeponides* sp. intermediate form – Widmark & Malmgren, p. 71, Pl. 2:4. □1990 *Neoeponides lunata* (Brotzen) – Thomas, p. 590, Pl. 2:5–6. □v1992a *Paralabamina* sp. intermediate form – Widmark & Malmgren, p. 112, Pl. 3:2.

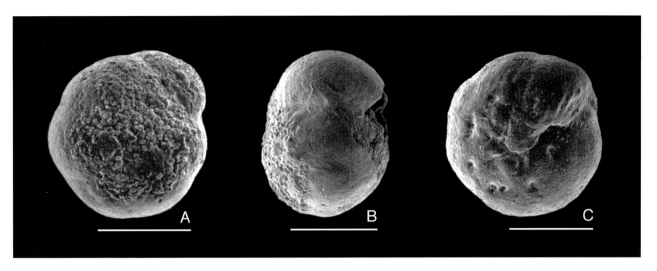

Fig. 40. Scale 100 µm. □A–C. *Scheibnerova*? sp. □A. Spiral view; sample 527-32-4, 68–69 cm (Maastrichtian). □B. Peripheral view; sample 527-32-4, 87–89 cm (Maastrichtian). □C. Umbilical view; sample 527-32-4, 87–89 cm (Maastrichtian).

Material. – 227 specimens.

Description. – Test trochospiral and equally to unequally biconvex; subcircular and lobated in outline. Periphery rounded to subangular. Walls calcareous and finely perforate. Aperture consisting of a simple arc at depression at base of last chamber, extending toward the umbilicus. Dorsal side (spiral side) convex and evolute. All chambers visible on dorsal side and arranged into three whorls. Dorsal sutures distinct, curved, and slightly depressed. Ventral side (umbilical side) convex and involute; umbilicus more or less closed. Only chambers of last whorl visible on ventral side; 6–7 in number. Ventral sutures distinct, radial, depressed, and slightly curved.

Occurrence. – Danian and Maastrichtian at Sites 525 and 527.

Remarks. – Specimens assigned to this taxon are morphologically intermediate between the specimens of *P. hillebrandti* and *P. lunata*. *Paralabamina* sp. intermediate form is biconvex, with a spiral side resembling that of *P. hillebrandti*, and an umbilical side resembling that of *P. lunata*. Tjalsma & Lohmann (1983) reported a species named *Neoeponides* cf. *lunata*, which shows close resemblance to *P.* sp. intermediate form.

Genus *Scheibnerova* Quilty, 1984

Type species and diagnosis. – See Loeblich & Tappan (1988, p. 642).

Scheibnerova? sp.

Fig. 40A–C

Synonymy. – □v1964 *Gyroidina globosa* (Hagenow) – Todd & Low, p. 412, Pl. 4:3. □1992a *Scheibnerova*? sp. – Widmark & Malmgren, p. 113, Pl. 6:2. □1992b *Scheibnerova*? sp. – Widmark & Malmgren, p. 402, Pl. 3:4–6.

Material. – Forty-three specimens.

Description. – Test trochospiral, subglobular, somewhat compressed dorsally–ventrally, and circular in outline. Periphery broadly rounded. Walls calcareous and perforate; most of the dorsal side covered with small knobs, which are absent on the ventral side; umbilical area covered by large pores forming small grooves and sockets. Aperture consisting of short interiomarginal slit; low apertural face. Dorsal side (spiral side) convex and evolute. Only chambers of final whorl visible on dorsal side; chambers small and low, 6–7 in number, increasing gradually in size as added. Dorsal sutures distinct in chambers of last whorl, depressed and oblique; sutures in previous whorls hardly visible. Ventral side (umbilical side) convex and involute; umbilicus closed by an extension of the last chamber. Only chambers of final whorl visible on the ventral side. Ventral sutures distinct, nearly straight, and slightly depressed.

Occurrence. – Maastrichtian at Sites 525 and 527.

Remarks. – Examination of the material of Todd & Low (1964) at the Smithsonian Institution led to the conclusion that they identified the same form under the name of *Gyroidina globosa* (Hagenow) [= *Gyroidinoides globosus* (Hagenow)]. In my opinion, *Scheibnerova*? sp. is distinctly different from *Gyroidinoides globosus* (Hagenow) in having very narrow chambers and a very low apertural face. In addition, *Scheibnerova*? sp. is covered with tiny knobs dorsally and large pores and grooves ventrally; these features are lacking in *Gyroidinoides globosus* (Hagenow). The status of this taxon is uncertain, and it is tentatively referred to the genus of *Scheibnerova* Quilty 1984. The specimens found here have two fundamental features in common with *S. protindica* Quilty (the type species of this genus), namely the tiny knobs dorsally and the large pores and grooves ventrally. However, *Scheibnerova*? sp. differs from the type species in having a broadly rounded periphery, instead of an acute one as in *S. protindica*.

Genus *Sliteria* Gawor-Biedowa, 1992

Type species. – Original designation by Gawor-Biedowa (1992, p. 155); *Sliteria varsoviensis*; Maastrichtian (Upper Cretaceous); Poland: Lublin Upland (borehole Tyszowce IG 1) – the Holy Cross Mountains region (boreholes Senisławice and Chwalibogowice 10S, depth 83,5 m), and in boreholes from southern Baltic Sea.

Diagnosis. – Original diagnosis of Gawor-Biedowa (1992): 'Test free, low trochospiral, ventral side involute, dorsal one semiinvolute; only final whorl visible. Ventral side with umbilicus obscured by short tongue-like umbilical flaps projecting from the umbilical margin of the chambers. Dorsal side with a depression in the center. Test margin rounded. Chambers on the ventral side coarsely perforate near the umbilicus. Sutures flat, nearly radial. Chambers with nodose ornamentation on the dorsa[l] side. Suture slightly raised and ended with spines near the depression in the central part of this side. Aperture interiomarginal with a narrow lip, extending also under tongue-like flaps on the ventral side. Test, smooth, finely perforated (with exception of large nearumbilical pores on ventral side). Wall and septa trilamellar. Outer and inner part of the wall is optically fibrous, inner one is prismatic.'

Sliteria varsoviensis Gawor-Biedowa, 1992

Fig. 41A–C

Synonymy. – □*1992 *Sliteria varsoviensis* sp. n. – Gawor-Biedowa, p. 156, Pl. 33:9–13. □v1994 *Gavelinella martini* (Sliter) – Speijer, p. 64, Pl. 3:1.

Material. – Sixty-five specimens.

Description. – Test trochospiral and subglobular. Periphery broadly rounded, somewhat truncated. Walls calcareous and perforated; dorsally rough, because of heavily spinose ornamentation; walls much smoother ventrally

than dorsally; small ridges between the sutures radiate from umbilical area toward periphery. Aperture hardly observable, seems to consist of an interiomarginal slit, extending from periphery to umbilicus. Dorsal side (spiral side) flat to weakly concave; not completely evolute. Chambers hardly visible on dorsal side, because of spinose ornamentation; dorsal sutures indistinct and slightly curved. Ventral side (umbilical side) convex and involute; umbilicus small and sometimes covered by a small umbilical flap. Only chambers of final whorl visible on ventral side. Ventral sutures distinct, depressed, curved, and limbate.

Occurrence. – Maastrichtian at Site 525.

Remarks. – This curious but morphologically stable form has been reported from El Kef (Tunisia) and the Negev and Sinai Deserts (Israel and Egypt, respectively) as *Gavelinella martini* (Sliter) by Speijer & Van der Zwaan (1996). *Gavelinella martini* was originally described as *Gyroidinoides quadratus martini* by Sliter (1968) from the Upper Cretaceaous of the southern California. Based on topotypic material kindly provided by W. Sliter, it is here concluded that *G. quadratus martini* differs from *S. varsoviensis* in being much flatter ventrally and less ornamented dorsally. The small radiating ridges on the umbilical side and the umbilical flaps of *Sliteria* are features that support its affinity to *Gavelinella* and *Stensioeina*, which are both here placed together with *Sliteria* in the subfamily Gavelinellinae, whereas Gawor-Biedowa (1992) originally placed *Sliteria* in the subfamily Gyroidinoidinae.

Genus *Stensioeina* Brotzen, 1936

Type species and diagnosis. – See Loeblich & Tappan (1988, p. 635).

Stensioeina pommerana Brotzen, 1940

Fig. 41D–F

Synonymy. – □v*1936 *Stensioeina pommerana* n.sp. – Brotzen, p. 166. □v1940 *Stensioeina americana* n.sp. – Cushman & Dorsey, p. 5, Pl. 1:7. □v1940 *Stensioeina labyrinthica* n.sp. – Cushman & Dorsey, p. 3, Pl. 1:5. □v1940 *Stensioeina pommerana* Brotzen – Cushman & Dorsey, p. 2, Pl. 1:4. □1992 *Stensioeina pommerana* Brotzen – Gawor-Biedowa, p. 161, Pl. 39:1–3. □v1994 *Stensioeina pommerana* Brotzen – Speijer, p. 62, Pl. 2:2.

Material. – Eight specimens.

Description. – Test trochospiral, planoconvex, and subcircular in outline. Periphery angular. Walls calcareous; dorsally coarsely perforated and rough, owing to ornamentation; ventrally perforated, final chamber coarsely

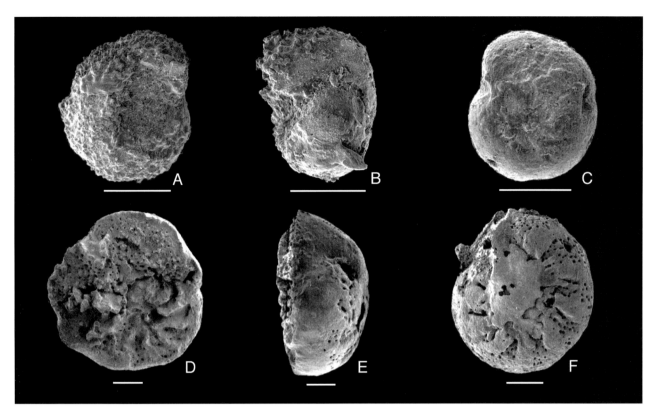

Fig. 41. Scale 100 µm. □A–C. *Sliteria varsoviensis* Gawor-Biedowa; sample 525A-40-3, 130–131 cm (Maastrichtian). □A. Spiral view. □B. Peripheral view. □C. Umbilical view. □D–F. *Stensioeina pommerana* Brotzen; sample 525A-40-3, 130–131 cm (Maastrichtian). □D. Spiral view. □E. Peripheral view. □F. Umbilical view.

perforated, small ridges between the sutures radiate from umbilical area toward periphery. Aperture consisting of an interiomarginal slit, extending from periphery to umbilicus. Dorsal side (spiral side) flat and evolute. All chambers visible on dorsal side and increase rapidly in size as added; about ten chambers in final whorl; last chamber in adult specimens extremely broad and narrow (length/breadth ratio about 1:3). Dorsal sutures indistinct, broad, and distinctly elevated, forming an irregular, reticulate pattern in adult (large) specimens. Ventral side (umbilical side) strongly convex and involute; umbilicus covered by white, opaque material, which is missing in some less-well preserved specimens. Only chambers of final whorl visible on ventral side. Ventral sutures distinct, depressed, curved, and limbate.

Occurrence. – Danian and Maastrichtian at Site 525.

Remarks. – Examination of a number of primary and secondary types filed under five different species names of *Stensioeina* at the Smithsonian Institution led to the conclusion that three (i.e. *S. americana*, *S. labyrinthica*, and *S. pommerana*) were intergrading morphologically, making a separation among these three impossible. *Stensioeina*

americana was the only type that was slightly different from the other two similar forms in being more or less concave dorsally, in having a more distinct elevated spiral suture, and by its smaller test size, features that, however, could vary within a population. In addition, a large number of *Stensioeina pommerana* Brotzen were available and examined at the Naturhistoriska Riksmuseet, Stockholm. These specimens are all within the *S. americana – S. labyrinthica – S. pommerana* complex, although the name *S. pommerana* is preferred for this taxon, because of its widespread use in the literature.

Family and genus indet.

'Rotaliid' sp.

Fig. 42A–C

Material. – Thirty specimens.

Description. – Test trochospiral, subglobular, and circular in outline. Periphery broadly rounded. Aperture difficult to distinguish but seems to consist of a short interiomar-

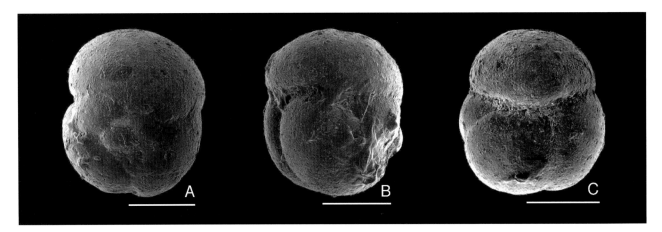

Fig. 42. Scale 100 μm. □A–C. 'Rotaliid'? sp.; sample 525A-40-2, 92–93 cm (Maastrichtian). □A. Spiral view. □B. Peripheral view. □C. Umbilical view.

ginal slit, located close to the umbilicus. Walls calcareous, finely perforate, and smooth. Dorsal side (spiral side) weakly convex and evolute with a distinct apex. Three highly inflated chambers in the final whorl dorsally, strongly suppressing the chambers of the previous whorls; chambers rapidly increasing in size as added; about four times longer than broad in spiral view. Dorsal sutures distinct, narrow, orthogonal to spiral suture, and depressed. Ventral side (umbilical side) strongly convex and involute; umbilicus open and shallow. Only chambers of final whorl visible ventrally. Ventral sutures distinct, broader than dorsal ones, straight, and depressed.

Occurrence. – Maastrichtian at Site 525.

Remarks. – The taxonomic status of this taxon is uncertain. Its triserial test shape is similar to that of triserial buliminids. However, the apertural features (interiomarginal slit) of this taxon do not fit the apertural definition of most triserial buliminids, where the aperture often consists of a loop-like opening extending from base of final chamber upward the apertural face.

References

Alth, A. 1850: Geognostisch-palaeontologische Beschreibung der nächsten Umgebung von Lemberg. *Naturwissenschaftliche Abhandlungen, Wien (1848–49) 3*, 171–284.

Barmawidjaja, D.M., Jorissen, F.J., Puskaric, S. & Van der Zwaan, G.J. 1992: Microhabitat selection by benthic foraminifera in the northern Adriatic Sea. *Journal of Foraminiferal Research 22*, 297–317.

Beckmann, J.P. 1978: Late Cretaceous smaller benthic foraminifers from Sites 363 and 364 DSDP Leg 40, southeast Atlantic Ocean. *In* Bolli, H.M., Ryan, W.B.F. *et al.* (eds.): *Initial Reports of the Deep Sea Drilling Project 40*, 759–782. U.S. Government Printing Office, Washington, D.C.

Bellagamba, M. & Coccioni, R. 1990: Deep-water agglutinated foraminifera from the Massignano section (Ancona, Italy), a proposed stratotype for the Eocene–Oligocene boundary. *In* Hemleben, C., Kaminski, M.A. & Scott, D.B. (eds.): *Paleoecology, Biostratigraphy, Pale-*

oceanography and Taxonomy of Agglutinated Foraminifera, 883–921. Kluwer, Dordrecht.

Berggren, W.A. & Aubert, J. 1975: Paleocene benthonic foraminiferal biostratigraphy, paleobiogeography and paleoecology of Atlantic–Tethyan regions: Midway-type fauna. *Palaeogeography, Palaeoclimatology, Palaeoecology 18*, 73–192.

Berggren, W.A. & Aubert, J. 1983: Paleogene benthonic foraminiferal biostratigraphy and paleobathymetry of the central Ranges of California. *U.S Geological Survey, Professional Paper 1213*, 4–21.

Berggren, W.A. & Miller, K.G. 1988: Paleogene tropical planktonic foraminiferal biostratigraphy and magnetobiochronology. *Micropaleontology 34*, 362–380.

Berggren, W.A., Kent, D.V., Flynn, J.J. 1985: Jurassic to Paleogene: Part 2, Paleogene geochronology and chronostratigraphy. *In* Snelling, N.J. (ed.): *The Chronology of the Geological Record. Geological Society of London, Memoir 10*, 141–195.

Boersma, A. 1984: Cretaceous-Tertiary planktonic foraminifers of the southeastern Atlantic, Walvis Ridge area, Deep Sea Drilling Project, Leg 74. *In* Moore, T.C., Rabinowitz, P.D. *et al.* (eds.): *Initial Reports of the Deep Sea Drilling Project 74*, 501–523. U.S. Government Printing Office, Washington, D.C.

Brotzen, F. 1936: Foraminiferen aus dem schwedischen untersten Senon von Eriksdal in Schonen. *Sveriges Geologiska Undersökning, Serie C 396*, 1–206.

Brotzen, F. 1948: The Swedish Paleocene and its foraminiferal fauna. *Sveriges Geologiska Undersökning C 493*, 1–140.

Caldeira, K. & Rampino, M.R. 1993: Aftermath of the End-Cretaceous mass extinction: possible biogeochemical stabilization of the carbon cycle and climate. *Paleoceanography 8*, 515–525.

Carsey, D.O. 1926: Foraminifera of the Cretaceous of central Texas. *University of Texas Bulletin 2612*, 1–53.

Charnock, M.A. & Jones, R.W. 1990: Agglutinated foraminifera from the Paleocene of the North Sea. *In* Hemleben, C., Kaminski, M.A. & Scott, D.B. (eds.): *Paleoecology, Biostratigraphy, Paleoceanography and Taxonomy of Agglutinated Foraminifera*, 139–244. Kluwer, Dordrecht.

Ciesielski, P.F., Sliter, W.V., Wind, F.H. & Wise, S.W. 1977: Paleoenvironmental analysis and correlation of a Cretaceous Islas Orcadas core from Falkland Plateau, southwest Atlantic. *Marine Micropaleontology 2*, 27–34.

Clark, M.W. & Wright, R.C. 1984: Paleogene abyssal foraminifers from the Cape and Angola Basins, South Atlantic Ocean. *In* Hsu, K.J., LaBrecque, J.L. *et al.* (eds.): *Initial Reports of the Deep Sea Drilling Project 73*, 459–480. U.S. Government Printing Office, Washington, D.C.

Cole, W.S. 1938: Stratigraphy and micropaleontology of two deep wells in Florida. *Florida Geological Survey, Bulletin 16*, 1–73.

Coccioni, R. & Galeotti, S. 1994: K–T boundary extinction: Geologically instantaeous or gradual event? Evidence from deep-sea benthic foraminifera. *Geology 22*, 779–782.

Corliss, B.H. 1981: Deep-sea benthonic foraminiferal fauna turnover near the Eocene/Oligocene boundary. *Marine Micropaleontology 6*, 367–384.

Corliss, B.H. 1985: Microhabitats of benthic foraminifera within deep-sea sediments. *Nature 314*, 435–438.

Corliss, B.H. 1991: Morphology and microhabitat preferences of benthic foraminifera from the northwest Atlantic Ocean. *Marine Micropaleontology 17*, 195–236.

Corliss, B.H. & Chen, C. 1988: Morphotype patterns of Norwegian Sea deep-sea benthic foraminifera and ecological implications. *Geology, 16*, 716

Corliss, B.H. & Emerson, S. 1990: Distribution of Rose Bengal stained deep-sea benthic foraminifera from the Nova Scotian continental margin and Gulf of Maine. *Deep-Sea Research 37*, 381–400.

Cushman, J.A. 1925: Some new foraminifera from the Velasco Shale of Mexico. *Contributions from the Cushman Laboratory for Foraminiferal Research 1*, 18–23.

Cushman, J.A. 1926a: Some foraminifera from the Mendez Shale of eastern Mexico. *Contributions from the Cushman Laboratory for Foraminiferal Research 2*, 16–26.

Cushman, J.A. 1926b: The foraminifera of the Velasco Shale of the Tampico Embayment. *Bulletin of American Association of Petroleum Geology 10*, 581–612.

Cushman, J.A. 1926c: *Eouvigerina*, a new genus from the Cretaceous. *Contributions from the Cushman Laboratory for Foraminiferal Research 2*, 3–6.

Cushman, J.A. 1927a: American upper Cretaceous species of Bolivina and related species. *Contributions from the Cushman Laboratory for Foraminiferal Research 2*, 85–91.

Cushman, J.A. 1927b: New and interesting foraminifera from Mexico and Texas. *Contributions from the Cushman Laboratory for Foraminiferal Research 3*, 111–117.

Cushman, J.A. 1931a: A preliminary report on the foraminifera of Tennessee. *Bulletin of the Tennessee Division of Geology 41*, 5–112.

Cushman, J.A. 1931b: Cretaceous foraminifera from Antigua, B.W.I. *Contributions from the Cushman Laboratory for Foraminiferal Research 7*, 33–46.

Cushman, J.A. 1931c: The foraminifera of the Saratoga Chalk. *Journal of Paleontology 5*, 297–315.

Cushman, J.A. 1933: New American Cretaceous foraminifera. *Contributions from the Cushman Laboratory for Foraminiferal Research 9*, 49–64.

Cushman, J.A. 1936: Cretaceous foraminifera of the family Chilostomellidae. *Contributions from the Cushman Laboratory for Foraminiferal Research 12*, 71–77.

Cushman, J.A. 1938: Additional new species of American Cretaceous foraminifera. *Contributions from the Cushman Laboratory for Foraminiferal Research 14*, 31–50.

Cushman, J.A. 1941: American Upper Cretaceous foraminifera belonging to *Robulus* and related genera. *Contributions from the Cushman Laboratory for Foraminiferal Research 17*, 55–69.

Cushman, J.A. 1944a: The foraminiferal fauna of the type locality of the Pecan Gap chalk. *Contributions from the Cushman Laboratory for Foraminiferal Research 20*, 1–17.

Cushman, J.A. 1944b: Foraminifera of the lower part of the Mooreville chalk of the Selma Group of Mississippi. *Contributions from the Cushman Laboratory for Foraminiferal Research 20*, 83–96.

Cushman, J.A. 1946: Upper Cretaceous foraminifera of the Gulf Coast region of the United States and adjacent areas. *U.S. Geological Survey, Professional Paper 206*, 1–241.

Cushman, J.A. & Campbell, A.S. 1935: Cretaceous foraminifera from the Moreno Shale of California. *Contributions from the Cushman Laboratory for Foraminiferal Research 11*, 65–73.

Cushman, J.A. & Church, C.C. 1929: Some Upper Cretaceous foraminifera from near Coalinga, California. *California Academy of Science, Proceedings, series 4, 18*, 497–530.

Cushman, J.A. & Deaderick, W.H. 1944: Cretaceous foraminifera from the Marlbrook Marl of Arkansas. *Journal of Paleontology 18*, 328–342.

Cushman, J.A. & Dorsey, A.L. 1940: The genus *Stensioeina* and its species. *Contributions from the Cushman Laboratory for Foraminiferal Research 16*, 1–6.

Cushman, J.A. & Hedberg, H.D. 1941: Upper Cretaceous foraminifera from Santander del Norte, Colombia, S.A. *Contributions from the Cushman Laboratory for Foraminiferal Research 17*, 79–100.

Cushman, J.A. & Jarvis, P.W. 1928: Cretaceous foraminifera from Trinidad. *Contributions from the Cushman Laboratory for Foraminiferal Research 4*, 85–103.

Cushman, J.A. & Jarvis, P.W. 1932: Upper Cretaceous foraminifera from Trinidad. *Proceedings of the United States National Museum 18*, 1–60.

Cushman, J.A. & Parker, F.L. 1935: Some American Cretaceous buliminas. *Contributions from the Cushman Laboratory for Foraminiferal Research 11*, 95–101.

Cushman, J.A. & Parker, F.L. 1936a: Notes on some Cretaceous species of Buliminella and Neobulimina. *Contributions from the Cushman Laboratory for Foraminiferal Research 12*, 5–10.

Cushman, J.A. & Parker, F.L. 1936b: Some American Eocene buliminas. *Contributions from the Cushman Laboratory for Foraminiferal Research 12*, 39–45.

Cushman, J.A. & Parker, F.L. 1940: New species of *Bulimina*. *Contributions from the Cushman Laboratory for Foraminiferal Research 16*, 44–48.

Cushman, J.A. & Renz, H.H. 1946: The foraminiferal fauna of the Lizard Springs Formation of Trinidad, British West Indies. *Cushman Laboratory for Foraminiferal Research, Special Publication 18*, 1–48.

Cushman, J.A. & Renz, H.H. 1947: Further notes on the Cretaceous foraminifera of Trinidad. *Contributions from the Cushman Laboratory for Foraminiferal Research 23*, 31–51.

Cushman, J.A. & Siegfus, S.S. 1935: New species of foraminifera from the Kreyenhagen Shale of Frenso County, California. *Contributions from the Cushman Laboratory for Foraminiferal Research 11*, 90–95.

Cushman, J.A. & Todd, R. 1943: The genus *Pullenia* and its species. *Contributions from the Cushman Laboratory for Foraminiferal Research 19*, 1–23.

Dailey, D.H. 1983: Late Cretaceous and Paleocene benthic foraminifers from Deep Sea Drilling Project Site 516, Rio Grand Rise, western South Atlantic. *In* Barker, P.F., Carlson, R.L., Johnson, D.A. *et al.* (eds.): *Initial Reports of the Deep Sea Drilling Project 72*, 757–782. U.S. Government Printing Office, Washington, D.C.

Dietz, R.S. & Holden, J.C. 1970: Reconstruction of Pangea: Break-up and dispersion of the continents, Permian to Recent. *Journal of Geophysical Reseach 75*, 4939–4956.

Dingle, R.V. & Simpson, E.S.W. 1976: The Walvis Ridge: A Review. *In* Drake, C.L. (ed.): *Geodynamics: Progress and Prospects*, 160–176. AGU Washington, D.C.

Douglas, R.G. 1973: Benthonic foraminiferal biostratigraphy in the Central North Pacific Leg 17, Deep Sea Drilling Project. *In* Winterer, E.L., Ewing, J.I. *et al.* (eds.): *Initial Reports of the Deep Sea Drilling Project 17*, 607–696. U.S. Government Printing Office, Washington, D.C.

Ewing, M. 1966: Crustal structure of the mid-ocean ridges: 4 – sediment distribution in the South Atlantic Ocean and the Cenozoic history of the Mid-Atlantic Ridge. *Journal of Geophysical Research 71*, 1611–1635.

Finlay, H.J. 1940: New Zealand foraminifera: Key species in stratigraphy 4. *Royal Society of New Zealand, Transactions and Proceedings 69*, 448–472.

Fisher, M.J. 1969: Benthonic Foraminifera from the Maestrichtian chalk of the Galicia Bank, west of Spain. *Palaeontology 12*, 189–200.

Frizzel, D.L. 1943: Upper Cretaceous foraminifera from nothwestern Peru. *Journal of Paleontology 17*, 331–353.

Futyan, A.I. 1976: Late Mesozoic and early Cainozoic benthic foraminifera from Jordan. *Palaentology 19*, 517–537.

Gawor-Biedowa, E. 1992: Campainian and Maastrichtian foraminifera from the Lublin Upland, eastern Poland. *Palaeontologia Polonica 52*, 1–187.

Geroch, S. & Nowak, W. 1984: A proposal of zonation for the Tithonian – late Eocene, based on arenaceous foraminifera from the Outer Catpathians. *In* Oertli, H. (ed.): *Benthos '83*, 225–239.

Gooday, A.J. 1986: Meiofaunal foraminiferans from the bathyal Porcupine Seabight (north-east Atlantic): size structure, standing stock, taxonomic composition, species diversity and vertical distribution in the sediment. *Deep-Sea Research 33*, 1345–1373.

Gooday, A.J. 1988: A response by benthic foraminifera to the deposition of phytodetritus in the deep sea. *Nature 332*, 70–73.

Gooday, A.J. 1994: The biology of deep-sea Foraminifera: A review of some advances and their applications in paleoceanography. *Palaios 9*, 14–31.

Gooday, A.J. & Lambshead, P.J.D. 1989: Influence of seasonally deposited phytodetritus on benthic foraminiferal populations in the bathyal northeast Atlantic: the species response. *Marine Ecology Progress Series 58*, 53–67.

Govindan, A. & Sastri, V.V. 1983: Upper Cretaceous arenaceous foraminifera from the Godavari-Krishna Basin, Andhra Pradesh, India. *In* Verdenius, J.V. Van Hinte, J.E. & Fortun, A.R. (eds.): *Proceedings of the First Workshop on Arenaceous Foraminifera, IKU Publication 108*, 33–55.

Grzybowski, J. 1896: Otwornice czerwonych iłów z Wadowic. [Foraminifera of the red clay of Wadowice.] *Rozprawy Academia umiejętności w Krakowie, Wydział matematyczno-przyrodniczy, Seria 2, 10*, 261–308. (In Polish.)

Grzybowski, J. 1898: Otwornice pokładów naftonośnych okolicy Krosna. [Foraminifera of the oil-bearing beds in the vicinity of Krosno.] *Rozprawy Academia umiejętności w Krakowie, Wydział matematyczno-przyrodniczy, Seria 2, 13*, 257–305. (In Polish.)

Hansen, H.J. 1970: Danian foraminifera from Nugssuaq, West Greenland. *Meddelelser om Grønland 193:2, Grønlands Geologiske Undersøgelse Bulletin 93*, 1–132.

Hagenow, F., von 1842: Monographie der Rügen'schen Kreide-Versteinerungen III: Mollusken. *Neues Jahrbuch fur Mineralogie, Geognosie, Geologie und Petrefakten-Kunde*, pp. 528–575. Stuttgart.

Heezen, B. & Tharp, M. 1965: Physiographic diagram of the Indian Ocean (with descriptive sheet). *Geological Society of America*. New York, N.Y.

Hart, M.B. 1983: Some thoughts on the ecology (and paleoecology) of the arenaceous foraminifera: a workshop report. *In* Verdenius, J.V. Van Hinte, J.E. & Fortun, A.R. (eds.): *Proceedings of the First Workshop on Arenaceous Foraminifera, IKU Publication 108*, 251–265.

Hermelin, J.O.R. & Malmgren, B.A. 1980: Multivariate analysis of environmentally controlled variation in *Lagena*: Late Maastrichtian, Sweden. *Cretaceous Research 1*, 193–206.

Hillebrandt, A., von 1962: Das Paleozän und seiner Foraminiferenfauna im Becken von Reichenhall und Salzburg. *Abhandlungen, Bayerische Akadademie der Wissenschaften, Mathematisch-Naturwissenschaftliche Klasse, N.F. 108*, 1–182.

Hiltermann, H. & Koch, W. 1950: Taxonomie und Vertikalverbreitung von Bolivinoides-Arten im Senon Nordwestdeutschlands. *Geologisches Jahrbuch 64*, 595–632.

Hurlbert, S.H. 1971: The nonconcept of species diversity: a critique and alternative parameters. *Ecology 52*, 577

Jones G.D. 1988: A paleoecological model of Late Paleocene 'Flysch-type' agglutinated foraminifera using the paleoslope transect approach, Viking Graben, North Sea. *Abhandlungen der Geologischen Bundesanstalt 41*, 143–153.

Jorissen, F.J., Barmawidjaja, D.M., Puskaric, S. & Van der Zwaan 1993: Vertical distribution of benthic foraminifera in the northern Adriatic Sea: the relation with the organic flux. *Marine Micropaleontology 19*, 131–146.

Kaminski, M.A. 1984: Shape variation in Spiroplectammina spectabilis (Grzybowski). *Acta Palaeontologica Polonica 29*, 29–49.

Kaminski, M.A. & Geroch, S. 1993: A revision of foraminiferal species in the Grzybowski Collection. *In* Kaminski, M.A. *et al.* (eds.): *The Origin of Applied Micropalaentology: the School of Jósef Grzybowski, Grzybowski Foundation, Special Publication 1*, 239–323.

Kaminski, M.A., Boersma, A., Tyszka, J. & Holbourn, A.E.L. 1995: Response of deep-water agglutinated foraminifera to dysoxic conditions in the California Boarderland basins. *In* Kaminski, M.A. Geroch, S. & Gasinski, M.A. (eds.): *Proceedings of the Fourth International Workshop on Agglutinated Foraminifera, Grzybowski Foundation, Special Publication 3*, 131–140.

Kaminski, M.A., Gradstein, F.M., Berggren, W.A., Geroch, S. & Beckmann, J.P. 1988: Flysch-type agglutinated assemblages from Trinidad: Taxonomy, stratigraphy and paleobathymetry. *Abhandlungen der Geologischen Bundesanstalt 41*, 155–227.

Kaiho, K. 1992: A low extinction rate of intermediate-water benthic foraminifera at the Cretaceous/Tertiary boundary. *Marine Micropaleontology 18*, 229–259.

Keller, G. 1988: Biotic turnover in benthic foraminifera across the Cretaceous/Tertiary boundary at El Kef. *Palaeogeography, Palaeoclimatology, Palaeoecology 66*, 153–171.

Keller, G. 1989a: Extended period of extinctions across the Cretaceous/Tertiary boundary in planktonic foraminifera of continental shelf sections: Implications for impact and volcanism theories. *Geological Society of America Bulletin 101*, 1408–1419.

Keller, G. 1989b: Extended Cretaceous/Tertiary boundary extinctions and delayed population change in planktonic foraminifera from Brazos River. *Paleoceanography 4*, 287–332.

Keller, G. 1992: Paleoecologic response of Tethyan benthic foraminifera to the Cretaceous–Tertiary boundary transition. *Studies in Benthic Foraminifera, Benthos '90*, 70–91. Tokai University Press, Tokai.

Kemton, R.A. 1979: The structure of species abundance and measurement of diversity. *Biometrics 35*, 307–321.

Klasz, I., de & Klasz, S., de 1990: Danian deep water (bathyal) agglutinated foraminifera from Bavaria and their comparison with approximately coeval agglutinated assemblages from Senegal and Trinidad. *In* Hemleben, C., Kaminski, M.A. & Scott, D.B. (eds.): *Paleoecology, Biostratigraphy, Paleoceanography and Taxonomy of Agglutinated Foraminifera*, 387–431. Kluwer, Dordrecht.

Klootwijk, C.T., Gee, J.S., Pierce, J.W., Smith, G.M. & McFadden, P.L. 1992: An early India–Asia contact: Paleomagnetic constraints from Ninetyeast Ridge, ODP Leg 121. *Geology 20*, 395–398.

Koutsoukos, E.A.M., Leary, P.N & Hart, M.B. 1990: Latest Cenomanian – earliest Turonian low-oxygen benthonic foraminifera: a case-study from the Sergipe basin (N.E. Brazil) and the western Anglo-Paris basin (southern England). *Palaeogeography, Palaeoclimatology, Palaeoecology 77*, 145–177.

Kuhnt, W. 1990: Agglutinated foraminifera of western Mediterranean Upper Cretaceous pelagic limestones (Umbrian Apennines, Italy & Betic Cordillera, southern Spain). *Micropaleontology 36*, 297–330.

Kuhnt, W. & Kaminski, M.A. 1993: Changes in community structure of deep water agglutinated foraminifers across the K/T boundary in the Basque Basin (northern Spain). *Revista Española de Micropaleontologia 25*, 57–92.

Kuhnt, W. & Moullade, M. 1991: Quantitative analysis of Upper Cretaceous abyssal agglutinated foraminiferal distribution in the North Atlantic – paleoceanographic implications. *Revue de Micropaléontologie 34*, 313–349.

Lamarck, J.B. 1804: Suite des mémoires sur les fossiles des environs de Paris. *Annales du Muséum Nationale d'histoire Naturelle, Paris 5*, 179–188.

Lewis, J., Watkins, H., Hartman, H. & Prinn, R. 1982: Chemical consequences of major impacts on Earth. *Geological Society of America, Special Paper 190*, 215–221.

Le Pichon, X. & Hayes, D.E. 1971: Marginal offsets, fracture zones & the early opening of the South Atlantic. *Journal of Geophysical Reseach 76*, 6283–6293.

Linke, P. & Lutze, G.F. 1993: Microhabitat preferences of benthic foraminifera: a static concept or a dynamic adaptation to optimize food acquisition? *Marine Micropaleontology 20*, 215–233.

Loeblich, A.R., Jr. & Tappan, H. 1964: Sarcodina, chiefly 'Thecamoebians' and Foraminiferida. *In* Moore, R.C. (ed.): *Treatise on Invertebrate Paleontology, Part C, Protista 2*, C1–C900. Geological Society of America and University of Kansas Press, Boulder, Col., and Lawrence, Kans.

Loeblich, A.R., Jr. & Tappan, H. 1988: *Foraminiferal Genera and Their Classification*. 970 pp. Van Nostrand Reinhold, New York, N.Y.

McGugan, A. 1964: Upper Cretaceous zone foraminifera, Vancover Island, British Columbia, Canada. *Journal of Paleontology 38*, 933–951.

Mackensen, A. & Douglas, R.G. 1989: Down-core distribution of live and dead deep-water benthic foraminifera in box cores from the Weddell Sea and the California continental borderland. *Deep-Sea Research 36*, 879–900.

MacLeod, N. & Keller, G. 1991a: Hiatus distributions and mass extinctions at the Cretaceous/Tertiary boundary. *Geology 19*, 497–501.

MacLeod, N. & Keller, G. 1991b: How complete are Cretaceous/Tertiary sections? A chronostratigraphic estimate based on graphic correlation. *Geological Society of America Bulletin 103*, 1439–1457.

McNeil, D.H. & Caldwell, W.G.E. 1981: Cretaceous rocks and their foraminifera in the Manitoba Escarpment. *The Geological Association of Canada, Special Paper 21*, 1–439.

Malata, E. & Oszczypko, N. 1990: Deep water agglutinated foraminiferal assemblages from Upper Cretaceous Red Shales of the Magura Nappe / Polish Outer Carpathians. *In* Hemleben, C., Kaminski, M.A. & Scott, D.B. (eds.): *Paleoecology, Biostratigraphy, Paleoceanography and Taxonomy of Agglutinated Foraminifera*, 507–524. Kluwer, Dordrecht.

Malmgren, B.A. & Sigaroodi, M.M. 1985, Standardization of species counts – the usefulness of Hurlbert's diversity-index in paleontology. *Bulletin of the Geological Institutions of the University of Uppsala, N.S. 10*, 111–114.

Marsson, T. 1878: Die Foraminiferen der weissen Schreibkreide der Insel Rügen. *Mitteilungen des Naturwissenschaftlichen Vereins fur Neu-Vorpommern und Rügen in Greifswald 10*, 115–196.

Martin, L.T. 1964: Upper Cretaceous and Lower Tertiary foraminifera from Fresno County, California. *Jahrbuch der Geologischen Bundesanstalt Wien, Sonderband 9*, 1–128.

Montgomery, H., Pessagno, E., Soegaard, K., Smith, C., Muñoz, I. & Passagno, J. 1992: Misconceptions concerning the Cretaceous/Tertiary boundary at the Brazos River, Falls County, Texas. *Earth and Planetary Science Letters 109*, 593–600.

Moore, T.C., Rabinowitz, P.D. *et al.* 1983: The Walvis Ridge transect, Deep Sea Drillig Project Leg 74: The geologic evolution of an oceanic plateau in the South Atlantic Ocean. *Geological Society of America Bulletin 94*, 907–925.

Moore, T.C., Rabinowitz, P.D. *et al.* 1984a: Site 525. *In* Moore, T.C., Rabinowitz, P.D. *et al.* (eds.): *Initial Reports of the Deep Sea Drilling Project 74*, 41–160. U.S. Government Printing Office, Washington, D.C.

Moore, T.C., Rabinowitz, P.D. *et al.* 1984b: Site 527. *In* Moore, T.C., Rabinowitz, P.D. *et al.* (eds.): *Initial Reports of the Deep Sea Drilling Project 74*, 237–306. U.S. Government Printing Office, Washington, D.C.

Moore, T.C., Rabinowitz, P.D. *et al.* 1984c: History of the Walvis Ridge. *In* Moore, T.C., Rabinowitz, P.D. *et al.* (eds.): *Initial Reports of the Deep Sea Drilling Project 74*, pp. 873–894. U.S. Government Printing Office, Washington, D.C.

Morgan, J.W. 1972: Deep mantle convection plumes and plate motions. *American Association of Petroleum Geology Bulletin 56*, 203–213.

Morkhoven, F.P.C.M., Van, Berggren, W.A. & Edwards, S.A. 1986: Cenozoic cosmopolitan deep-water benthic Foraminifera. *Bulletin Centres Recherches Exploration-Production Elf-Aquitaine, Mémoires 11*, 1–412. Pau, France.

Morrow, A.L. 1934: Foraminifera and ostracoda from the Upper Cretaceous of Kansas. *Journal of Paleontology 8*, 186–205.

Nakkady, S.E. 1959: Biostratigraphy of the Um Elghanayem section, Egypt. *Micropaleontology 5*, 453–472.

Nomura, R. 1991: Paleoceanography of upper Maestrichtian to Eocene benthic foraminiferal asemblages at Sites 752, 753 and 754, eastern Indian Ocean. *In* Weissel, J. Peirce, J. Taylor, E., Alt, J. *et al.* (eds.): *Proceedings ODP, Scientific Results 121*, 3–29. College Station, TX (Ocean Drilling Program).

Nuttall, W.L.F. 1930: Eocene foraminifera from Mexico. *Journal of Paleontology 4*, 271–293.

Nyong, E.E. & Olsson, R.K. 1984: A paleoslope model of Campanian to Lower Maestrichtian foraminifera in the North American Basin and adjacent continental margin. *Marine Micropaleontology 8*, 437–477.

d'Orbigny, A.D. 1840: Mémoire sur les foraminifères de la craie blanche du bassin de Paris. *Société Géologique de France, Mémoires 4*, 1–51.

Parker, W.C., Clark, M.W., Wright, R.C. & Clark, R.K. 1984: Population dynamics, Paleogene abyssal benthic foraminifers, eastern South Atlantic. *In* Hsü, K.J., LaBrecque, J.L. *et al.* (eds.): *Initial Reports of the Deep Sea Drilling Project 73*, 481–486. U.S. Government Printing Office, Washington, D.C.

Pospichal, J.J. 1995: Calcareous nannofossil biostratigraphy and abundance changes across the K–T boundary of northeastern Mexico. *GSA Abstracts with Programs, Annual Meeting 1995 27*, A347.

Proto Decima, F. & Bolli, H.M. 1978: Southeast Atlantic DSDP Leg 40 Paleogene benthic foraminifers. *In* H.M. Bolli, W.B.F. Ryan *et al.* (eds.): *Initial Reports of the Deep Sea Drilling Project 40*, pp. 787–809. U.S. Government Printing Office, Washington, D.C.

Prinn, R.G. & Fegley, B., Jr. 1987: Boline impacts, acid rain & biospheric traumas at the Cretaceous–Tertiary boundary. *Earth and Planetary Science Letters 83*, 1–15.

Quilty, P.G. 1984: Cretaceous foraminiferids from Exmouth Plateau and Kerguelen Ridge, Indian Ocean. *Alcheringa 8*, 225–241.

Rathburn, A.E. & Corliss, B.H. 1994: The ecology of living (stained) benthic foraminifera from the Sulu Sea. *Paleoceanography 9*, 87–150.

Reiss, Z. 1954: Upper Cretaceous and Lower Tertiary Bolivinoides from Israel. *Contributions from the Cushman Foundation for Foraminiferal Research 5*, 154–164.

Reuss, A.E. 1845: *Die Versteinerungen der böhmischen Kreideformation 1*. 58 pp. Schweizerbart, Stuttgart.

Reuss, A.E. 1846: *Die Versteinerungen der böhmischen Kreideformation: (Nachtrage und Erganzungen zur ersten Abteilung) 2*. 148 pp. Schweizerbart, Stuttgart.

Reuss, A.E. 1851: Über die fossilen Foraminiferen und Entomostraceen der Septarienthone der Umgegend von Berlin. *Zeitschrift der Deutschen Geologischen Gesellschaft 3*, 49–92.

Reuss, A.E. 1860: Die Foraminiferen der westphälischen Kreideformation. *Sitzungsberichte der K. Academie der Wissenschaften in Wien, Mathematische-Naturwissenschaftliche Classe 40*, 147–238.

Reuss, A.E. 1862: Der Foraminiferen des norddeutschen Hils und Gault. *Sitzungsberichte der K. Academie der Wissenschaften in Wien, Mathematische-Naturwissenschaftliche Classe 46*, 5–100.

Rosoff, D.B. & Corliss, B.H. 1992: An analysis of Recent deep-sea benthic foraminiferal morphotypes from the Norwegian and Greenland Seas. *Palaeogeography, Palaeoclimatology, Palaeoecology 91*, 13–20.

Rudolph, E., Olsson, R.K., Habib, D. & Lui, C. 1995: Integrated dinoflagellate and planktonic foraminiferal evidence of the K/T boundary event in Texas, Alabama, Georgia and Mexico. *GSA Abstracts with Programs, Annual Meeting 1995 27*, A348.

Said, R. & Kenawy, A. 1956: Upper Cretaceous and Lower Tertiary foraminifera from northern Sinai, Egypt. *Micropaleontology 2*, 105–173.

Saint-Marc, P. 1987: Biostratigraphic paleoenvironmental study of Paleocene benthic and planktonic foraminifers, Site 605, Deep Sea Drilling Project Leg 93. *In* Van Hinte, J.E., Wise, S.W. *et al.* (eds.): *Initial Reports of the Deep Sea Drilling Project 93*, 539–547. U.S. Government Printing Office, Washington, D.C.

Scheibnerová, V. 1971: Implications of deep sea drilling in the Atlantic for studies in Australia and New Zealand – some new views on Cretaceous and Cainozoic palaeogeography and biostratigraphy. *Search 2*, 251–254.

Scheibnerová, V. 1972: Some new views on Cretaceous biostratigraphy, based on the concept of foraminiferal biogeoprovinces. *Records of the Geological Survey of New South Wales 14*, 85–87.

Scheibnerová, V. 1973: Non-tropical Cretaceous foraminifera in Atlantic deep-sea cores and thier implications for continental drift and palaeooceanography of the South Atlantic Ocean. *Records of the Geological Survey of New South Wales 5*, 19–46.

Scheibnerová, V. 1978: Aptian and Albian benthic foraminifers of Leg 40, Sites 363 and 364, southern Atlantic. *In* H.M. Bolli, W.B.F. Ryan *et al.* (eds.): *Initial Reports of the Deep Sea Drilling Project 40*, pp. 741–757. U.S. Government Printing Office, Washington, D.C.

Shackleton, N.J. *et al.* 1984: Accumulation rates in Leg 74 sediments. *In* Moore, T.C., Rabinowitz, P.D. *et al.* (eds.): *Initial Reports of the Deep Sea Drilling Project 74*, pp. 621–644. U.S. Government Printing Office, Washington, D.C.

Shaffer, F.R. 1984: The origin of the Walvis Ridge: sediment/basalt compensation during crustal separation. *Palaeogeography, Palaeoclimatology, Palaeoecology 45*, 87–100.

Sliter, W.V. 1968: Upper Cretaceous foramonifera from southern California and northwestern Baja California, Mexico. *Kansas University Paleontological Contribution 49*, 1–141.

Sliter, W.V. 1976: Cretaceous foraminifers from the southwestern Atlantic Ocean, Leg 36. *In* Barker, P., Dalziel, I.W.D. *et al.* (eds.): *Initial Reports of the Deep Sea Drilling Project 36*, 519–573. U.S. Government Printing Office, Washington, D.C.

Sliter, W.V. 1977: Cretaceous benthic foraminifers from the western South Atlantic Leg 39, Deep Sea Drilling Project. *In* Supko, P.R., Perch-Nilsen, K. *et al.* (eds.): *Initial Reports of the Deep Sea Drilling Project 39*, 657–697. U.S. Government Printing Office, Washington, D.C.

Smit, J. 1990: Meteorite impact, extinctions and the Cretaceous–Tertiary boundary. *Geologie en Mijnbouw 69*, 187–204.

Smit, J., Alvarez, W., Claeys, S., Montanari, S. & Roep, Th.B. 1994: Misunderstandings regarding the KT boundary deposits in the Gulf of Mexico. *In*: *New Developments Regarding the KT Event and Other Catastrophes in Earth History*, 116–117. *LPI Contribution 825*. Lunar and Planetary Institute, Houston, Tex.

Smit, J., Alvarez, W. & Claeys, S. 1995: Tsunami induced sandstone beds at the KT boundary around the Gulf of Mexico: Consequences of the Chicxulub impact. *GSA Abstracts with Programs, Annual Meeting 1995 27*, A347.

Speijer, R.P. 1994: Extinction and recovery patterns in benthic foraminiferal paleocommunities across the Cretaceous/Paleogene and Paleocene/Eocene boundaries. *Geologica Ultraiectina 124*, 1–191.

Speijer, R.P. & Van der Zwaan, G.J. 1996: Extinction and survivorship of southern Tethyan benthic foraminifera across the Cretaceous/Paleogene boundary. In Hart, M.B. (ed.): *Biotic Recovery from Mass Extinction Events*, 242–371. *Geological Society, Special Publication 102*.

Thiede, J., Vallier, T.L. *et al.* 1981: Site 465. *In* Thiede, J., Vallier, T.L. *et al.* (eds.): *Initial Reports of the Deep Sea Drilling Project 62*, 199–282. U.S. Government Printing Office, Washington, D.C.

Thomas, E. 1990: Late Cretaceous through Neogene deep-sea benthic foraminifers (Maud Rise, Weddell Sea, Antarctica). *In* Barker, P.F.,

Kennett, J.P. *et al.* (eds.): *Proceedings of the Ocean Drilling Program, Scientific Results 113B*, 571–594. College Station, TX (Ocean Drilling Program).

Thomas, E. & Vincent, E. 1988: Early to middle Miocene deep-sea benthic foraminifera in the equatorial Pacific. *Revue de Paleobiologie, Special Volume (Benthos '86) 2*, 583–588.

Tjalsma, R.C. & Lohmann, G.P. 1983: Paleocene–Eocene bathyal and abyssal benthic Foraminifera from the Atlantic Ocean. *Micropaleontology, Special Publication 4*, 1–90.

Todd, R: 1970: Maestrichtian (Late Cretaceous) foraminifera from a deep-sea core of southwestern Africa. *Revista Española de Micropaleontología 2*, 131–154.

Todd, R. & Low, D. 1964: Cenomanian (Cretaceous) foraminifera from the Puerto Rico Trench. *Deep-Sea Research 11*, 395–414.

Trujillo, E.F. 1960: Upper Cretaceous foraminifera from near Redding, Shasta County, California. *Journal of Paleontology 34*, 290–346.

Vasilenko, V.P. 1950: Paleotsenovye foraminifery tsentral'noj chasti Dnjepro-Donetskogo bassejna. [Paleocene foraminifera of the central part of the Dnjepr-Donez basin.] In: *Mikrofauna SSSR*, 177–224. *Trudy VNIGRI 4*. (In Russian.)

Verdenius, J.G. & Hinte, J.E. 1983: Central Norwegian – Greenland Sea: Tertiary arenaceous foraminifera, biostratigraphy and environment. *In* Verdenius, J.V., Van Hinte, J.E. & Fortun, A.R. (eds.): *Proceedings of the First Workshop on Arenaceous Foraminifera, IKU Publication 108*, 173–223.

Walker, G. & Jacob, E. 1798: *In* Kanmacher, F. (ed.): *Addam's Essays on the Microscope, 2nd Ed*. Dillon & Keating, London.

Webb, P.N. 1973: Upper Cretaceous–Paleocene Foraminifera from Site 208 (Lord Howe Rise, Tasman Sea), DSDP, Leg 21. *In* Burns, R.E. Andrews, J.E. *et al.* (eds.): *Initial Reports of the Deep Sea Drilling Project 21*, 541–573. U.S. Government Printing Office, Washington, D.C.

White, M.P. 1928a: Some index Foraminifera of the Tampico embayment area of Mexico: Part I. *Journal of Paleontology 2*, 177–215.

White, M.P. 1928b: Some index Foraminifera of the Tampico embayment area of Mexico: Part II. *Journal of Paleontology 2*, 280–317.

White, M.P. 1929: Some index Foraminifera of the Tampico embayment area of Mexico: Part III. *Journal of Paleontology 3*, 30–58.

Widmark, J.G.V. & Malmgren, B.A. 1988: Differential dissolution of Upper Cretaceous deep-sea benthonic foraminifers from the Angola Basin, South Atlantic Ocean. *Marine Micropaleontology 13*, 47–78.

Widmark, J.G.V. & Malmgren, B.A. 1992a: Benthic foraminiferal changes across the Cretaceous–Tertiary boundary in the deep sea; DSDP Sites 525, 527 and 465. *Journal of Foraminiferal Research 22*, 81–113.

Widmark, J.G.V. & Malmgren, B.A. 1992b: Biogeography of terminal Cretaceous deep-sea benthic foraminifera from the Atlantic and Pacific Oceans. *Palaeogeography, Palaeoclimatology, Palaeoecology 92*, 375–405.

Wilson, J.T. 1965: Submarine fracture zones, aseismic ridges and the ICSU line: proposed western margin of the East Pacific Rise. *Nature 207*, 907–911.

Zakharow, M.V. 1981: Geophysical characteristics and internal structure of the Walvis Ridge region (Southeast Atlantic). *Geotectonics 15*, 443–450.

Appendix 1

List of primary and secondary types examined at the Cushman Collection, Smithsonian Institution, Washington, D.C., USA. The types are arranged in alphabetical order under the names given on the slides. The abbreviations 'CC:' and 'USNM:' denote the 'Cushman Collection No.' and 'U.S. National Museum No.', respectively.

Primary types

Allomorphina minuta Cushman, 1936, holotype (CC: 23516; Cushman 1936, p. 72, Pl. 13:3).

Allomorphina navarroana Cushman, 1936, holotype (CC: 23520; Cushman 1936, p. 73, Pl. 13:1).

Anomalina velascoensis Cushman, 1925, holotype (CC: 4345; Cushman 1925, p. 21, Pl. 3:3).

Astigerina crassaformis Cushman & Siegfus, 1935, holotype (CC: 22358; Cushman & Siegfus 1935, p. 94, Pl. 14:10), paratypes (CC: 22359; Cushman & Siegfus 1935, p. 94).

Bolivina incrassata Reuss var. *limonensis* Cushman, 1926, holotype (CC: 5068; Cushman 1926a, p. 19, Pl. 2:2).

Bolivina rhomboidea Cushman, 1926, holotype (CC: 5039; Cushman 1926a, p. 19, Pl. 2:3.

Bolivinoides decorata (Jones) var. *delicatula* Cushman, 1927, holotype (CC: 5139; Cushman 1927a, p. 90, Pl. 12:8).

Bolivinopsis minuta Said & Kenawy, 1956, holotype (USNM: P 4067; Said & Kenawy 1956, p. 138, Pl. 3:27).

Bolivitina eleyi Cushman, 1927, holotype (CC: 5552; Cushman 1927a, p. 91, Pl. 12:11).

Bolivitina planata Cushman, 1927, holotype (CC: 6701; Cushman 1927b, p. 115, Pl. 23:9).

Bulimina arkadelphiana var. *midwayensis* Cushman & Parker, 1936, holotype (CC: 23136; Cushman & Parker 1936b, p. 42, Pl. 7:9).

Bulimina exigua Cushman & Parker, 1935, holotype (CC: 22371; Cushman & Parker 1935, p. 99, Pl. 15:7).

Bulimina incisa Cushman, 1926, holotype (CC: 5157; Cushman 1926b, p. 592, Pl. 17:9).

Bulimina kickapooensis Cole, 1938, paratype (CC: 369367; Cole 1938, p. 45).

Bulimina pectinata Cushman & Parker, 1940, holotype (CC: 32761; Cushman & Parker 1940, p. 45, Pl. 8:20).

Bulimina petroleana Cushman & Hedberg var. *spinea* Cushman & Renz, 1946, holotype (CC: 46706; Cushman & Renz 1946, p. 37, Pl. 6:13).

Bulimina rudita Cushman & Parker, 1936, paratypes (CC: 32720; Cushman & Parker 1936b, p. 45).

Bulimina spinata Cushman & Campbell, 1935, holotype (CC: 22947; Cushman & Campbell 1935, p. 72, Pl. 11:11).

Bulimina taylorensis Cushman & Parker, 1935, holotype (CC: 22367; Cushman & Parker 1935, p. 96, Pl. 15:3).

Bulimina triangularis Cushman & Parker, 1935, holotype (CC:22369; Cushman & Parker 1935, p. 97, Pl. 15:4).

Bulimina trinitatensis Cushman & Jarvis, 1928, holotype (CC: 9682; Cushman & Jarvis 1928, p. 102, Pl. 14:12).

Buliminella beaumonti Cushman & Renz, 1946, holotype (CC: 46694; Cushman & Renz 1946, p. 36, Pl. 6:7), paratypes (CC: 46695; Cushman & Renz 1946, p. 36).

Buliminella carseyae Plummer var. *plana* Cushman & Parker, 1936, holotype (CC: 22571; Cushman & Parker 1936a, p. 8, Pl. 2:7).

Buliminella vitrea Cushman & Parker, 1936, holotype (CC: 22575; Cushman & Parker 1936a, p. 7, Pl. 2:4).

Clavulina amorpha Cushman, 1926, holotype (CC: 5153; Cushman 1926b, p. 589, Pl. 17:5).

Clavulina trilatera Cushman var. *aspera* Cushman, 1926, holotype (CC: 5154; Cushman 1926b, p. 589, Pl. 17:3).

Clavulina trilatera Cushman, 1926, holotype (CC: 5155; Cushman 1926b, p. 588, Pl. 17:2).

Eouvigerina aculeata Cushman, 1933, holotype (CC: 19047; Cushman 1933, p. 62, Pl. 7:8).

Eouvigerina americana Cushman, 1926, holotype (CC: 4986; Cushman 1926c, p. 4, Pl. 1:1).

Eponides bollii Cushman & Renz, 1946, holotype (CC: 46770; Cushman & Renz 1946, p. 44, Pl. 7:23), paratypes (CC: 46771; Cushman & Renz 1946, p. 44).

Eponides bronnimanni Cushman & Renz, 1946, holotype (CC: 46774; Cushman & Renz 1946, p. 45, Pl. 7:24), paratypes (CC: 46775; Cushman & Renz 1946, p. 45).

Eponides mariei Said & Kenawy, 1956, holotype (USNM: P 4142; Said & Kenawy 1956, p. 148, Pl. 5:2).

Eponides sigali Said & Kenawy, 1956, holotype (USNM: P 4144; Said & Kenawy 1956, p. 148, Pl. 5:6).

Gaudryina velascoensis Cushman, 1925, holotype (CC: 4349; Cushman 1925, p. 20, Pl. 8:7).

Gyroidina infrafossa Finlay, 1939, holotype (USNM: 689013; Finlay 1940, p. 462, Pl. 66:181-183).

Gyroidina reussi Said & Kenawy, 1956, holotype (USNM: P 4149; Said & Kenawy 1956, p. 149, Pl. 5:10).

Gyroidinoides imitata Olsson, 1960, holotype (USNM: 626464; Olsson 1960, p. 36, Pl. 6:2-4).

Pseudouvigerina plummerae Cushman, 1927, holotype (CC: 6700; Cushman 1927b, p. 115, Pl. 23:8).

Pullenia cretacea Cushman, 1936, holotype (CC: 23527; Cushman 1936, p. 75, Pl. 13:8); paratypes (CC: 23528; Cushman 1936, p. 75).

Pullenia jarvisi Cushman, 1936, holotype (CC: 15459; Cushman 1936, p. 77, Pl. 13:6.

Pullenia reussi Cushman & Todd, 1943, holotype (CC: 39053; Cushman & Todd 1943, p. 4, Pl. 10:13).

Pulvinulinella texana Cushman, 1938, holotype (CC: 34903; Cushman 1938, p. 49, Pl. 8:8).

Stensioeina americana Cushman & Dorsey, 1940, holotype (CC: 35124; Cushman & Dorsey 1940, p. 5, Pl. 1:7).

Stensioeina labyrinthica Cushman & Dorsey, 1940, holotype (CC: 35112; Cushman & Dorsey 1940, p. 3, Pl. 1:5).

Truncatulina excolata Cushman, 1926, holotype (CC: 5104; Cushman 1926a, p. 22, Pl. 3:2).

Truncatulina spinea Cushman, 1926, holotype (CC: 5083; Cushman 1926a, p. 22, Pl. 2:10).

Truncatulina velascoensis Cushman, 1925, holotype (CC: 4344; Cushman 1925, p. 20, Pl. 3:2).

Secondary types

Anomalina beccariiformis (White, 1928), plesiotype (CC: 46805; Cushman & Renz 1946, p. 48, Pl. 21:22); plesiotype (CC: 47361; Cushman & Renz 1947, p. 50, Pl. 12:14).

Bulimina reussi Morrow, 1934, plesiotype (CC: 22433; Cushman & Parker 1935, p. 99, Pl. 15:8); plesiotype (CC: 37991; Cushman & Hedberg 1941, p. 95, Pl. 22:30); plesiotype (CC: 42071; Cushman & Deadrick 1944, p. 337, Pl. 53:6);

Bulimina trinitatensis Cushman & Jarvis, 1928, plesiotype (CC: 46698; Cushman & Renz 1946, p. 37, Pl. 6, figs. 8,9).

Conorbina marginata Brotzen, 1936, Sec. types (USNM: P 54; from lowest Senonian, Eriksdal, Sweden, identified by Brotzen).

Conorbina marginata Brotzen, 1936, syntype (USNM: P 3094; from lowest Senonian, Eriksdal II, Sweden, identified by Brotzen).

Globorotalites conicus (Carsey, 1926), hypotype (USNM: P 4137; Said & Kenawy 1956, p. 147, Pl. 4:43).

Globorotalia multisepta Brotzen, 1936, syntypes (USNM: P 48; from Lower Senonian, Eriksdal, Sweden, identified by Brotzen).

Globorotalites multiseptus (Brotzen, 1936), hypotype (USNM: 643591; Todd 1970, p. 146, Pl. 3:4).

Globorotalites micheliniana (d'Orbigny, 1840), Sec. types (lacking collection no.; from lower Upper Campanian, Bavaria, Germany; identified by Hagn).

Gyroidina beisseli White, 1928, Sec. types [CC: 46769; (no further information on slide)].

Gyroidina depressa (Alth, 1850), plesiotype (CC: 40245; Cushman 1944a, p. 13, Pl. 3:2); plesiotype (CC: 42429; Cushman 1944b, p. 95, Pl. 14:23).

Gyroidina girardana (Reuss, 1851), plesiotype (CC: 46767; Cushman & Renz 1946, p. 44, Pl. 7:20); plesiotype (CC: 42506; Cushman 1944b, p. 95, Pl. 14:24); hypotype (USNM: 643147; Nakkady 1959, p. 460, Pl. 3:1).

Gyroidina globosa (Hagenow, 1842), plesiotype (CC: 15625; Cushman 1931c, p. 310, Pl. 35:19); plesiotype (CC: 15446; Cushman & Jarvis 1932, p. 47, Pl. 14, figs. 4-5); plesiotype (CC: 40247; Cushman 1944a, p. 13, Pl. 3:3); plesiotype (CC: 46764; Cushman & Renz 1946, p. 44, Pl. 7:15); hypotypes (USNM: P 4147; Said & Kenawy 1956, p. 149, Pl. 5:5); topotypes (CC: 13220; CC: 13167; CC: 13170; from Senonian, Campanian, Island of Ruegen, Germany).

Gyroidina nitida Reuss, 1845, topotype [CC: 3154; from the Upper Cretaceous (Turonian) 'Plänarmergel' (Dresden, Germany)].

Lenticulina rotulata (Lamarck, 1804), plesiotype (CC: 28354; Cushman 1941, p. 67, Pl. 16:13); plesiotype (CC: 28357; Cushman 1946, p. 56, Pl. 19:3); plesiotype (CC: 28355; Cushman 1946, p. 56, Pl. 19:5).

Pullenia coryelli White, 1929, plesiotype (CC: 41981; Cushman & Deadrick 1944, p. 339, Pl. 53:26); plesiotype (CC: 46786; Cushman & Renz 1946, p. 47, Pl. 8:9).

Pulvinulinella florealis (White, 1928), Sec. types (CC:382166; recieved from Hofker).

Pulvinulinella? florealis (White, 1928), plesiotype (CC: 46781; Cushman & Renz 1946, p. 46, Pl. 8:4-5); Sec. types [CC: 46782; (no further information on slide)].

Pulvinulinella texana Cushman, 1938, plesiotype (CC: 40254; Cushman 1944a, p. 14, Pl. 3:5); Sec. types [CC: 40252; (no further information on slide)].

Quadratobuliminella pyramidalis de Klaz, 1953, topotype (USNM: 433512; Loeblich & Tappan 1964, p. C546, Pl. 442:2).

Robulus macrodiscus (Reuss, 1862), plesiotype (CC: 15341; Cushman & Jarvis 1932, p. 23, Pl. 7:3); Sec. types [CC: 47962 & CC: 61975 (no further information on slide)].

Rotalia depressa Alth, 1850, topotype (CC: 16245; from Cretaceous, Mucronata Kreide, Lemberg, Germany).

Rotalia nitida Reuss, 1845, topotype (CC: 3154; from Turonian, 'Plänarmergel', Dresden, Germany).

Stensioeina exsculpta (Reuss, 1860), topotypes [CC: 19810, CC: 35142, CC: 35143; (no further information on slide)]; topotype (USNM: 433611; Loeblich & Tappan 1964, p. C763, Pl. 627:7).

Stensioeina pommerana Brotzen, 1936, plesiotype (CC: 35118; Cushman & Dorsey 1940, p. 2, Pl. 1:4).

Valvulina allomorphinoides Reuss, 1860, plesiotype (CC: 15242; Cushman 1931a, p. 53, Pl. 9:6); plesiotype (CC: 15492; Cushman 1931b, p. 43, Pl. 6:2); plesiotype (CC: 15440; Cushman & Jarvis 1932, p. 46, Pl. 13:17); plesiotype (CC: 38001; Cushman & Hedberg 1941, p. 96, Pl. 23:9); plesiotype (CC: 46758; Cushman & Renz 1946, p. 44, Pl. 7:13-14).

Appendix 2

Absolute abundance of taxa encountered at Sites 525 and 527 arranged according to their last occurrence. Figures at top of colums refer to the 'Sample No.' given in Table 2; the K–Pg boundary is located between sample 4 and 5, and 24 and 25 in Holes 525A and 527, respectively (see Table 2).

Site 525	1	2	3	4	5	6	7	8	9	10	11	12	13	14	15	16
Bolivinoides paleocenicus	3			3												
Oridorsalis? sp.		9	2	3												
Marginulina spp.			1													
Aragonia spp.	8	6	8	8		2	9									
Nuttallinella florealis	6	1	8	13		1	2									
Marssonella? sp. triserial form	2	2	2	2			2									
Bulimina trinitatensis	23	17	21	40	1		12	1								
Pullenia coryelli	4	4	14	6	1	7	12	6								
Gavelinella spp. lobated forms	2	3		10		1				1						
Valvalabamina sp. involute	5	2	1	10		3	2		8			2				
Gyroidinoides beisseli	3	6	1	6	2	1	7				1	2				
Buliminella beaumonti	4	3	6			1	3	6	2			2				
Bolivinoides delicatulus	6	5	1	2			2					1				
Guttulina spp.	1		1			1	1	3				1				
Angulogavelinella avnimelechi	6	6	9	5	3	5	5	3	3	5				1		
Patellina? spp.	2		2	2		2	1		1			1		1		
Bulimina spinea	4	5	6	9	8	1	1		1	1		1			1	
Gavelinella beccariiformis	74	85	51	92	72	58	57	38	38	54	60	52	52	58	37	98
Nuttallides truempyi	57	75	34	48	50	55	64	13	29	10	12	46	60	10	22	38
Cibicidoides hyphalus	26	32	21	50	38	38	43	29	24	20	17	31	18	27	30	33
Gaudryina pyramidata	5	1			4	18	12	17	28	18	16	22	4	19	17	4
Nuttallinella sp. A	3	1		3	8	9	14	20	19	12	8	7	13	3		6
Tritaxia spp.	11	12	10	12	6	7	13	7	4	2	4	4	1	7	7	5
Gyroidinoides spp.	4	5	3	13	2	7	7	6	16	6	6	5	6	6	11	5
Cibicidoides velascoensis	10	12	23	20	9	6	7	6	2		2		3			2
Pullenia sp. compressed form	3	4	2	3	3	9	4	8	11	8	13	10	3	5	9	5
Paralabamina lunata	3	3		2	1	7	4	14	7	10	8	2	2	12	11	2
Osangularia velascoensis	11	14	17	16	3	5	8	1	3	2	2	1			2	3
Gyroidinoides tellburmaensis	9	8	1	10		4	7	5	7	3	3	6	2	3	5	6

Site 525	1	2	3	4	5	6	7	8	9	10	11	12	13	14	15	16
Spiroplectammina spectabilis	3		2	5			1		2		1	7	3			1
Lenticulina cf. *macrodisca*	3	3	5	3	3	4	7	2	2	2	4	10	5	6	5	12
Osangularia texana	1	7	8	9	2	4	8	4	4	2	5	9	3	2	4	4
Osangularia? spp. juvenile(?)	8	4	13	15	1	3	5	3		1				1	2	16
Paralabamina sp. interm. form	10	24	4	10		3	4		1							3
Gyroidinoides goudkoffi	1		2	1		3	2	5	5	7	10	4			4	10
Dorothia pupa	5	8	3	5	1	1	8	5		2	6	2		1	1	3
Lenticulina cf. *rotulata*	6	4	3	11	5	1	5	2		1	4	2	1	1	2	2
Brizalina incrassata	2				4	2	1		1	3	4	4	3	3		20
Spiroplectammina spp. (juv.)	1					5	3	7	6	8	4	3		1	3	1
Oridorsalis? complex	1				3	4	4	3	4	3	1		2	1	5	7
Pyramidina sp.	5	7	4	3	1	3	3	1	1	1				3	3	1
Quadratobuliminella spp.	1	1		1	1	2	3	9	4		3	1	1	2	1	1
Globulina cf. *lacrima*	1		1	2	2	2	1	2	8	3	2	1	1	2		2
Lagena cf. *sulcata*	2	2		3	1	3	2	1	1		3	4	1		2	1
Valvalabamina? praeacuta	7	4	1	6												8
Marssonella oxycona	5	2	1	5	1	1	1	1	1		1	1			1	1
Cibicidoides dayi	2	2		2		2	3	1	1		1	2	2			3
Lagena sp. globous form	2	2	2		1	1	4			2	1	1		1	2	1
Lagena sp. ovate form	1	1	2		1	1	1		1	1		3			1	1
Ammodiscus cretaceus	1					1	1	1		2	2			2	1	3
Spiroplectammina spp. (calc.)	4	1		1	2		1									1
Nodosaria cf. *limbata*	2			2					2							2
Hyperammina-Bathysiphon spp.	X	X	X	X	X	X	X	X	X	X	X	X	X	X	X	X
Alabamina sp. C	2			2			1		1					1		
Bolivinoides cf. *clavatus*		1							1							
Nonionella spp.		1	1			3	1		2		4	1		2		
Quadrimorphina allomorphinoides		1				4		3	5	3	15	4	1	1	2	
Dorothia bulletta		1	1	2	1	1	1	2	2		4	1	1		1	
Gyroidinoides nitidus		2		1	1	2	1	4	5	7	9	8	5	4	3	2
Cibicides cf. *excavatus*			2						2			4				
Lagena sp. hispid form				1							1	1	3			
Neoflabellina spp.				1				1			1			3		
Stensioeina pommerana				1				1				1			5	
Astacolus spp.			2	1	1	1	2	5	7	3	3	2	4	4	3	3
Paralabamina hillebrandti			1	1			2	1	2		2				3	2
Bolivinoides cf. *postulatus*				1				1								
Valvalabamina sp. evolute				1		2	2	3	2							
Cibicidoides sp. plano-convex				4	4	2	3	2	10	1	6	10	2	6		
Lenticulina spp.				8	3	1	5	3	3	1		1		1	5	
Nonion spp.				1			4	1	2	4		5		1	5	
Buliminella cf. *beaumonti* / *B.* cf. *plana*				3	6	12	7	7	11	7	9	16	14	29	22	12
Gyroidinoides globosus				1	2	2										6
Nodosaria velascoensis				1	1		2	1			2	1		1		1
Allomorphina cf. *navarroana*					2											
Saracenaria spp.					1	2	1	2		2	1	2				
Gavelinella? sp. conical form					1		1	4		1		2				
Quadrimorphina camerata					1						2	4				
Ellipsoidella? spp.					1	1	1	1	1			1				
Goesella rugosa					1	1	1	3	2	2	2	2	2	2		
Spiroplectammina laevis					4	5		6	7	1	6	3		5	3	
Pleurostomella spp.					2			2	1				1		4	
Pyrolina? spp.					1		1		2			2			2	
Eouvigerina subsculptura					20	31	24	40	93	62	97	76	33	129	127	32
Praebulimina spp./*P. reussi*					12	12	5	13	29	15	23	26	5	18	130	3
Bolivinoides draco					3						1		1	37	21	9
Pseudouvigerina plummerae					1	1	2		5	9	2	3	1	2	13	19
Spiroplectammina israelskyi					9	1	6	4	7	4	10			8	5	3
Alabamina sp. A					3		1	10	6	5	6	1	1	8	8	5
Gyroidinoides quadratus					2	1								24	8	5
Bolivinoides decoratus					5	2	1	1						3	1	10
Ellipsogladulina? spp.					1											
Osangularia cordierana					1	4										
Globulina horrida					1			1	2			1				
Tritaxia aspera					1	4	9	9	9	9	21	3	7	4	20	
Rosalina? sp.					2	2	3	2							1	
Ellipsodimorphina? spp.					1		1								1	

Site 525	1	2	3	4	5	6	7	8	9	10	11	12	13	14	15	16
Loxostomum sp.							1	3	6	3	3	7	2	3		
Laevidentalina spp.							1	1	2		4	3	1	2	3	
Scheibnerova? sp.							1				1			1	2	2
Reussella szajnochae								1	1		2					
Fursenkoina? spp.								1			1					
Pullenia jarvisi								2			2	3				
'Rotaliid' sp.								3	7	2	7	3	2	6		
Sliteria varsoviensis								1	19	11	1	2			15	16
Fissurina spp.								1							3	
Glomospira? spp.								1		1	4	1		1	5	1
Reussoolina apiculata									1							
Pandaglandulina? spp.									1		2					
Ellipsopolymorphina? spp.									1	1		1				
Globorotalites cf. *conicus*									3	4	5	1		4	2	2
Nodogenerina spp.										1						
Osangularia sp. lenticulate form											3		1			
Allomorphina minuta											1					
Nodosarella spp.												1			3	
Globorotalites sp. C												1				16
Citharina? sp.																1
Taxa identified	52	48	48	56	51	60	68	56	53	44	47	56	33	41	43	44
Unid. mioliniids										1	3					
Unid. agglutinated	1	12	2	15	5	7	8	7	12	3	8	7	1	4	16	4
Unid. calcareous	41	25	22	46	19	19	31	19	42	12	25	18	17	18	24	24
Total number of specimens	413	436	330	560	350	414	497	404	560	368	499	479	295	522	657	471

Site 527	17	18	19	20	21	22	23	24	25	26	27	28	29	30	31	32	33	34	35	36
Nuttallinella florealis	1	3	1	3	3	1														
Pyramidina sp.	2		1	1	1		1													
Nodosarella spp.			3																	
Nodogenerina spp.			4		2	1														
Alabamina sp. C				1																
Bulimina trinitatensis					1															
Fursenkoina? spp.					3															
Ellipsobulimina spp.								1												
Nodosaria velascoensis								1												
Pullenia cf. *cretacea*	2	8	10	20	25	24	8	10	2											
Lagena sp. ovate form	1	3	3	1	4	5	6	3	3	1										
Gavelinella? sp. conical form	2	6	5	3	2	3		3			1									
Nonionella spp.	1		1	3	1		1						1							
Fissurina spp.	1		2	1	3	5	2	2	3				1	1						
Gyroidinoides sp. C	4	8	6	4	8	21	4	10	27	4	5	9	9	1						
Aragonia spp.	3	1			3	1	1	2	2		1	2			1					
Lagena sp. globous form	1	1	4	1	6	5	8	2	1			1	1	1	1					
Nodosaria cf. *limbata*	1		2		1	1			1							1				
Osangularia velascoensis	1	1		1				6	1					1		1				
Lenticulina spp.	1			1		5			2	1						2	1			
Laevidentalina spp.	1	1	1	1	3	1	2	3	6		1	2	1	4	1	2	2	2		
Osangularia? spp. juvenile	3	3	3	3	5	6	4		2			1					1		1	
Nuttallides sp. b	21	6	14	8	21	14	12	13	1		5	2		4					1	
Spiroplectammina dentata	2	7	5	5	6	4	3	1			1	3		1				1	5	
Valvalabamina sp. evolute	1	5	8	4	12	4	3	1	10	1	3	1	3	4	4	8	6	1	5	
Marssonella oxycona	2	3	2	1	8	2	1	3	2	1	2	1	2		2	2			1	
Gavelinella beccariiformis	22	14	26	41	44	46	5		118	46	18	8	7	11	17	6	29		10	36
Nuttallides truempyi	13	29	34	17	19	51	34	24	21	21	19	23	21	11	20	23	40	17	22	26
Valvalabamina sp. involute	11	8	25	21	34	51	31	15	2			1	2							2
Tritaxia spp.	14	32	44	29	16	1	2		6	6	4	2	8	5	7	2	13		2	5
Nuttallides sp. a	1	5	9	2	32	32	17	32	2	2	3	6	2	2						19
Gyroidinoides beisseli	8		6	9	24	9	3	9	10	6	10	9	10	4	10		17		5	4
Oridorsalis? complex	4	6	2	5	9	11	3	7	12	4	3	3	2	7	7	8	4	7	6	9
Gyroidinoides spp.	1	3	10	6	12	7	7	4	21	8	1	1	9	2		3	5	5		6
Globorotalites sp. B	1	3	7				1		10	8	6	2	5	2	2	1	11	2	2	2
Alabamina sp. a	2	5	5	1	2	2	1	1		4		5	2	5	2	2	5	3	3	3
Spiroplectammina spectabilis	6	1	5	3	1	10	1	3	4	1	2	1			3	1	3		2	5
Oridorsalis? sp.	4		2	2	5	2			4	4		1			1	2	3	8	6	1

Site 527	17	18	19	20	21	22	23	24	25	26	27	28	29	30	31	32	33	34	35	36
Lenticulina cf. *rotulata*	7	2	2	3	6	5		1	3	2	1		2	2		4	2	1		1
Globulina cf. *lacrima*	2	2	4	5	4	6	5	6	1	2			1				2			1
Pullenia sp. compressed form	2	1	5	3	4	1	1	8	6					4	1		1			2
Nonion spp.	1	1	4		1	7	1	1					1		2		2			2
Globorotalites sp. C	1											2	2		7		6			2
Glomospira? spp.	1			2								4	1	4			1			1
Cibicidoides velascoensis	2		2	1	1							1				2	1			2
Hyperammina–Bathysiphon spp.	X	X	X	X	X	X	X	X	X	X	X	X	X	X	X	X	X	X	X	X
Globorotalites sp. A		1	1				1		1											
Pyrolinoides? spp.	1		3	1	1	1							1							
Quadrimorphina allomorphinoides		2	3	2	10	15	5	55	3				1				1			
Buliminella cf. *beaumonti* / *B.* cf. *plana*		1							11	5	9	7	1	3	1		5			
Marssonella? sp. triserial form		2			6	4	1		5	2	4	1		2	2	1	1			
Globulina horrida		1	2	1	2	3	3	1	1	2			1					1		
Bulimina spinea		3	3	1	5	3	7	6	4	1	4	1			1	2	5		2	
Osangularia texana		3	7	5	10	3	3									1	1		3	
Paralabamina sp. intermediate form		4	11	10	21	15	17	29	12	2		5	4	3	2	4	18	4	2	5
Gyroidinoides quadratus		1			11	6			4	3		4	4	8	4	2		7	6	3
Valvalabamina? praeacuta		3		4	3	7		4	5	3		2	1	4		3	3		2	4
Paralabamina lunata		2			6	1		6	6	3	2		2	3	1	1				4
Astacolus spp.		2				1	1	2	1	1			1			3	2		1	3
Gavelinella spp. lobated forms		1			5			2	4	1				1					2	1
Vaginulina sp.			1		1	2			1				1							
Spiroplectammina spp. (juveniles)			1	1	3	1	1	2		1		2	1	1	2	2				
Tobolia? spp.			1	2	4	3		1	1						2		3			
Pleurostomella spp.			1					1	1		2				1	1	1			
Ellipsogladulina? spp.			1					1					1			1	2			
Ellipsopolymorphina? spp.			1		1	1								1			1			
Pullenia jarvisi			1									2						1		
Guttulina spp.			2	2		4	2	1	3	2	4	2			2		2	3	2	
Paralabamina hillebrandti			2	3	4	2	1	2	12	8	8	2			7	4	9		3	9
Allomorphina trochoides			1						1		1					1				
Dorothia bulletta			4									1						2		
Reussella szajnochae			1							2		2	1		4	1	1	6		2
Pullenia coryelli			2																	2
Ellipsodimorphina spp.				1							1									
Bolivinoides cf. *clavatus*				1					1	3		1	4	1	14			2		
Ammodiscus cretaceus				2					1	1		2		1			1	3		
Osangularia cordierana				1		2			3	3	1	2	1		1	8	6	4	3	
Cibicidoides hyphalus				1				1	24	13	6	9	10	16	9	13	22	12	15	15
Praebulimina spp./*P. reussi*				1	1			1	12	7	8	2	7	4	3		16	7	1	6
Rosalina? sp.				14	2		1	2	2				1		1					1
Lenticulina cf. *macrodisca*				1				1	1		2		2	1		2	3	2	2	
Quadrimorphina camerata				3	2		4	2	2	2										2
Ellipsoidella? spp.				1				1		2										1
Brizalina incrassata				1												1				2
Valvalabamina sp. lobated form					3	1										1				
Pyrolina? spp.					1	1			2	1		1	1	1			1			
Lagena cf. *sulcata*					1				1							1	2			
Osangularia sp. lenticulate form							3		5	1	2			2						5
Nuttallinella sp. A								1	4	1	7	7	8		4	8				
Gyroidinoides globosus								1		2							1			
Spiroplectammina laevis								1			1						1		1	
Nodosaria aspera									1					1						
Allomorphina minuta									2	1		1					1			
Scheibnerova? sp.									7	9	5	2	1	1		2	7		2	
Gaudryina pyramidata									9	3	4	5	1	2	3	4	2	2	2	4
Goesella rugosa									4	3			4	1	4	3	2	1	3	5
Bulimina velascoensis										1	4	3								
Bolivinoides cf. *postulatus*									3	3	4	1	1							
Dorothia pupa									1	1	1	2	2	3						
Gyroidinoides goudkoffi									3					1	1			1		
Buliminella beaumonti									2	1		1		3					1	1
Spiroplectammina israelskyi											2		1	1	2	3	1			2
Angulogavelinella avnimelechi												1		1						4
Gyroidinoides nitidus														1						1
Gyroidinoides tellburmaensis														1						

Site 527	17	18	19	20	21	22	23	24	25	26	27	28	29	30	31	32	33	34	35	36
Quadratobuliminella spp.															2				1	
Spiroplectammina spp. (calcareous)															1					
Globorotalites spineus															1					
Cibicidoides dayi																1				
Number of taxa identified	39	43	52	48	66	56	46	51	64	52	42	55	50	46	49	43	51	27	34	42
Unid. mioliniids																	1			
Unid. agglutinated	1	3	3	3	10	1		1	5	1	2	2	3	3	6	4	2	3	2	3
Unid. calcareous	9	13	21	12	24	34	14	32	64	21	7	2	6	14	27	16	14	5	5	19
Total number of specimens	164	210	331	266	489	458	239	328	509	240	178	170	167	156	208	163	298	108	132	234